Heal the Beasts

WORKS BY PHILIPP SCHOTT

THE ACCIDENTAL VETERINARIAN SERIES

The Accidental Veterinarian:
Tales from a Pet Practice

How to Examine a Wolverine:
More Tales from the Accidental Veterinarian

The Battle Cry of the Siamese Kitten:
Even More Tales from the Accidental Veterinarian

DR. BANNERMAN VET MYSTERIES

Fifty-Four Pigs:
A Dr. Bannerman Vet Mystery (#1)

Six Ostriches:
A Dr. Bannerman Vet Mystery (#2)

Eleven Huskies:
A Dr. Bannerman Vet Mystery (#3)

OTHER

The Willow Wren:
A Novel

Heal the Beasts

A JAUNT THROUGH THE CURIOUS HISTORY OF THE VETERINARY ARTS

Philipp Schott DVM

Published by ECW Press
665 Gerrard Street East
Toronto, Ontario, Canada M4M 1Y2
416-694-3348 / info@ecwpress.com

Cover design: Jessica Albert
Cover artwork: *Livre de chasse*, Gaston Phébus, fol. 40v
Author photo: Marlon Evan

LIBRARY AND ARCHIVES CANADA CATALOGUING IN PUBLICATION

Title: Heal the beasts : a jaunt through the curious history of the veterinary arts / Philipp Schott, DVM.

Names: Schott, Philipp, author

Description: Includes bibliographical references.

Identifiers: Canadiana (print) 20250140381 | Canadiana (ebook) 2025014039X

ISBN 9781770417830 (softcover)
ISBN 9781778523960 (ePub)
ISBN 9781778523977 (PDF)

Subjects: LCSH: Veterinary medicine—History. | LCSH: Veterinarians—History.

Classification: LCC SF615 .S36 2025 | DDC 636.089—dc23

This book is funded in part by the Government of Canada. *Ce livre est financé en partie par le gouvernement du Canada.* We acknowledge the support of the Canada Council for the Arts. *Nous remercions le Conseil des arts du Canada de son soutien.* We would like to acknowledge the funding support of the Ontario Arts Council (OAC) and the Government of Ontario for their support. We also acknowledge the support of the Government of Ontario through the Ontario Book Publishing Tax Credit, and through Ontario Creates.

Canada Council for the Arts Conseil des arts du Canada

PRINTED AND BOUND IN CANADA

PRINTING: FRIESENS 5 4 3 2 1

This book is dedicated to my half a million or so colleagues worldwide.
We are far too humble about what we have accomplished.

Contents

Preface

Cards on the table. I am not a professional historian. I am not even much of an amateur one. But I have a passion for history, a passion for veterinary medicine, and a passion for storytelling. The book in your hands, or your ears, is the product of the confluence of these three passions. As you'll discover, it's perhaps an idiosyncratic telling of veterinary medicine's stories across history. While it is intended to inform, and hopefully entertain, it's not intended to be authoritative or comprehensive. This is not a textbook. I know I run the risk of irritating some of my colleagues with my omissions or with the perspectives I take, but it is a risk I am fully comfortable with.

This book is intended for multiple audiences. It's intended for the general reader because I think every profession is misunderstood: Teachers, farmers, lawyers, clowns — all misunderstood. Veterinarians too. And it's intended for my fellow vets because I think every profession misunderstands itself. We are usually too deeply immersed in the minutiae of what is immediately before us to take a broader view of ourselves. Understanding where we came from is helpful.

I also want to use this obscure niche topic — the history of veterinary medicine — to tell a bigger story. It is often the case that the history of one thing mirrors the history of many things. In recent years, we've seen histories appear for salt, screws, tulips, marine chronometers, bananas, and the colour indigo. Each of these illuminated something more. So, why not veterinary medicine? If I do my job well, you will come away not only with a specific understanding of why and how mankind has attempted to treat ill and injured animals in the past, but also with a general understanding of the evolution of empathy in our society. When we heal the beasts, we often heal something larger.

CHAPTER ONE

Waterloo

When I'm in a major gallery, I rush by these paintings on my way to the Picassos or the Monets or the washrooms. I expect you do too. I'm talking about those massive 18th- and 19th-century canvases depicting battle scenes, often from the Napoleonic Wars but, depending on where you are, possibly the American War of Independence or the Seven Years' War. They're just too lurid. Too grimly realistic. They're not appealing to the modern eye or even especially interesting unless you're a military history buff. But next time you're there, stop for a moment and take a careful look. Usually, there are swarms of men in those bright red uniforms we associate with old-fashioned tin toy soldiers and nutcrackers. Usually, they are screaming or looking alarmed or resolute. They're almost always brandishing bright sabres or long muskets. There's smoke everywhere and churned mud and death. There's a lot to look at. And often there are horses.

Next time you're in front of one of these works, look at the horses' eyes and mouths. Historically, painters depicted horses in battle paintings as just part of the equipment. The artist may have viewed them as noble, but they were typically generic and emotionally inert figures. However, from the mid-18th century onwards, painters gave horses individuality and depicted clear signs of suffering: their eyes wide and white with terror, their nostrils flared, their mouths open and gasping as they fight for their lives while struggling to carry their masters deeper into battle.

Twelve thousand horses died in one day at the Battle of Waterloo in 1815.* Many wandered the battlefield, gravely injured. This description of the aftermath of the battle from the Waterloo Association's website brings the horror vividly to life. (You may want to skip this if you are of a sensitive disposition.)

> But perhaps the horses called forth even greater pity from those that witnessed their terrible suffering. The horses were often mutilated by cannonballs, tearing out their intestines, which they dragged around behind them until their strength failed them. Many more had legs torn away causing them to patiently sit or lay upon the ground, whilst chewing away at the grass within reach; their mournful eyes silently imploring someone to finish them off. The most awful of all according to eyewitnesses, were those horses that had the lower portion of their heads ripped away, few could look at these horrors impassively.

This appalling mass slaughter of horses in battle prompted alarm in war ministries throughout Europe, at how expensive it was to lose so many valuable animals, at how difficult it was to replace them. Cannons and muskets could be manufactured in a matter of days, but foals could not be sent directly to war. Replacement takes time — gestation, growth, training. In an age where cavalry reigned supreme on the battlefield, the loss of so many horses was a serious practical worry. The 18th century

---

* Other sources say seven thousand. Either way, it's a lot. The estimates for people killed vary even more widely, with numbers ranging from ten to thirty thousand. This speaks to the hellish chaos of the battle and to the haphazard record-keeping of the times. A gruesome postscript is that the massive number of horse and human bones at the battlefield site became the foundation of a brisk fertilizer trade as they were good sources of calcium and phosphorus.

had featured almost continuous warfare, so the assumption was that this would go on into the 19th century. Horses needed to be saved whenever possible.

Other forces that would eventually give birth to the modern veterinary profession were already at work. For example, several decades earlier, a cattle plague had devastated herds throughout Europe, threatening a food supply already made tenuous by the wars. This led to the foundation on August 4, 1761, of the first veterinary school in the world, in Lyon, France. Practical concerns about the medical care of horses for war and cattle for eating were two threads that wove together with others and led, ultimately, to the foundation of the veterinary profession as we understand it today.

One of these threads was the dawning of a more modern conception of empathy towards animals. The suffering of horses emerged as a topic of great concern among the middle and upper classes of Western Europe and Great Britain. Deeply empathic relationships with animals had existed at various times in history in various cultures around the world, but this was the start of a sustained change in the West. Only a little over a century and a half prior, in 1637, René Descartes infamously declared that animals were automata, mere meat machines, whose cries when injured were no different than the screech of metal on metal in a foundry. Descartes's opinion was commonly held and used to defend vivisection, the live dissection of an unanaesthetized animal to demonstrate the anatomy in the living state. This is unimaginably cruel and pointless to the modern mind but was commonplace at the time and seen as perfectly rational.

(It's interesting to note that Descartes had a little dog named Monsieur Grat, who apparently accompanied him everywhere and whom the philosopher spoke to and was visibly affectionate towards. Did Descartes make an exception for Monsieur Grat? Was he truly unmoved if his little companion cried out? I doubt it. Human philosophies are subject to change through history, but human emotions, less so. And humans have always been subject to hypocrisy and self-serving rationalizations.)

By the 1760s, the dominant philosophy was shifting, as we can see in Voltaire's *Treatise on Tolerance*: "It still seems to me that one must never have observed animals in order not to distinguish among them the different voices of need, suffering, joy, fear, love, anger, and all their affections; it would be strange if they expressed so well what they wouldn't feel." It was the Age of Enlightenment, and what we can call the "Circle of Concern" was finally beginning to expand. For millennia prior, only wealthy and powerful men were deemed worthy of society's full benefits, attentions, and concerns. Women, with a few notable exceptions, lay outside this Circle of Concern, as did children, foreigners, the poor, the different, the unfortunate. Animals were so far beyond the edge of the Circle of Concern that no reasonable person could foresee their inclusion at any point in the future, no matter how distant. It was impossible.

But as the circle expanded, the impossible suddenly became possible. To be more accurate, this circle has never been perfectly round and crisply defined. It protruded in various directions, with these protrusions differing in how well they covered their areas. Sometimes the protrusions retracted again for a time. It's really more of an amoeba or a blob than a circle, but "Amoeba of Concern" sounds like the title of a bulletin from the U.S. Centers for Disease Control and Prevention, and "Blob of Concern," an issuance from NASA. Regardless — circle, amoeba, or blob — the living beings that society truly cares about have dramatically increased in number, with the biggest expansion occurring because of the Enlightenment. The process is ongoing today.

But I'm getting ahead of myself. Far ahead. Before we attend to the Enlightenment Era birth of modern Western veterinary medicine and follow its growth and maturation through to the present day, I need to take you back to 12,205 BCE and introduce you to a puppy. Her name is Bonnie.

CHAPTER TWO

The Old Woman

The plants hadn't been easy to find, but her knowledge had not failed her. At this time of year, the best place was on the far edge of the wood, just before the marsh. Here the willows were dense, dense enough to discourage others from trying to forage there. But the old woman knew how to move through the willows to the spots where the plants grew in secret.

She hated the marsh, though. That's where the illness came from. She was sure of it. As the warmth and humidity of summer grew, the stinking vapours rose and seeped through the wood to the village. The vapours had killed her daughter and her daughter's man. Dead five days now. Sometimes death came from curses uttered by the people across the great river, and sometimes it was due to the disfavour of the Gods. But sometimes it was bad air or bad water or bad food. This time it was the foul marsh air.

She was heartsick at not having guessed it properly when her daughter died, but now that the little dog was ill, she knew. Their enemies would not curse a dog, nor would a dog ever raise the ire of the Gods. Before approaching the marsh edge, the old woman had chanted the words of protection and spilled some of her blood on the earth. She did this because it was the correct process, honed over countless generations, but she wasn't worried. Her own mother had often told the story of how, as a little girl, the old woman had become very ill from the fumes but survived. She told her that this survival was the best protection of all. She was weak in some ways, but strong in this one.

The dog's signs were different — green discharge from the nose, tremors, watery stool — whereas her daughter and her daughter's man had become burning hot and coughed blood. However, the timing could not be coincidental, and she knew from experience that marsh air illness took many different forms. The stink had been subtle before, but it was obvious now.

The old woman walked back to the shelter as quickly as she could. The puppy wagged her tail when she approached. The puppy was too weak to walk very far, but she could still wag that tail like she had all the energy in the world. The old woman stroked her and smiled. The little dog had been her daughter's favourite. Most dogs were trained to be hunters or guards, but this one had been destined to be more of a companion, especially since all her daughter's children had died. The puppy was very smart, even smarter than some of the old woman's people, such as her useless brother. And she was very gentle with her tiny sharp teeth, not like some puppies that you had to cuff across the snout and toss out of the shelter. This puppy was all the old woman had to remember her daughter by. The rest of the people had moved to higher ground for the hot time, but she couldn't bear to leave her daughter's gravesite.

The old woman went to her grinding stone at the entrance to the shelter and pounded the plants together into a paste. There were three of them. She was careful to get the proportions just right and to put them in in the right sequence with just a small amount of water at the end. Her mother had taught her this; she had learned it from her mother, and so on through all the mothers, going back to when the people first came here. Done correctly and given at the right time, it helped with the marsh stinking sicknesses, for people and for dogs.

The puppy kept wagging her tail, although more slowly, as the old woman scraped the paste onto the roof of the puppy's mouth. She then got her to swallow some water and offered her a piece of deer meat. The puppy didn't want the meat, but otherwise she looked a little better, if only in the brightness of her brown eyes. When this was done, the old woman picked up the puppy and carried her over to a tree stump near the grave, just beyond the shelter, overlooking the great river. The two of them sat there together for a long time as the sun slid below the hills to the west and the river turned silver, then grey, then black.

Dogs were the first animal to be domesticated. The curiously precise date of 12,205 BCE mentioned at the end of the previous chapter comes from carbon dating the bones of the earliest dog found beside human remains. Man's best friend indeed. I suspect that there are error bars of a few centuries on either side of this year, but let's go with it as it's kind of fun to have a specific date to mark as the birth of the human-animal bond. The bones were found in a riverside suburb of Bonn, Germany, called Oberkassel, so the dog is known to science as the Bonn-Oberkassel dog. I'll call her Bonnie for short.

Archaeologists are amazing people. Not only have they been able to date these bones, but they have also been able to say with confidence that this dog was afflicted with distemper virus in three separate bouts between the ages of 19 and 23 weeks, and then died at the age of 28 weeks.* So Bonnie was a puppy. A sick puppy. Moreover, based on the size of her teeth, the archaeologists know that Bonnie was a dog pup, not a wolf pup.

Typically, teeth become smaller through the domestication process. This is one of the things that distinguishes domestication from taming. People can tame individual wild animals, but their progeny will be just as wild. Take tigers used in circuses, for example, or bears used in movies. With a lot of effort they can be trained not to eat their handlers, but there has

* The remains were first found in 1914, but it was a 2018 study published in the *Journal of Archaeological Science* by the Dutch veterinarian Luc Janssens and his team of archaeologists that determined that Bonnie had distemper. So, although I praise archaeologists at the beginning of the paragraph to be polite, veterinarians deserve some praise as well.

been no intent to change the genetics of any offspring they might have. Domestication implies a deliberate process of selecting the calmest and easiest-to-handle animals and breeding them together so that each generation becomes just a little less likely to eat you, even without training. In a famous experiment in Russia, silver foxes were bred with this in mind, and after only a few generations changes were already apparent in the pups. By the 40th generation, the fox colony was fully domesticated.* Interestingly, they also ended up looking far more dog-like, with floppy ears, multicoloured coats, and curly tails, because these traits are genetically linked to tameness in canid species.

But back to Bonnie.

Bonnie was found with the remains of two adult humans: a slender woman in her early twenties and a beefier man of around 40. All three skeletons had been covered with a reddish iron oxide powder that archaeologists theorize was used in a ritual burial, as opposed to these bones just randomly ending up there and then randomly being coated in rust dust. So, our Bonnie was beloved and cared for. One could argue that ancient humans were often also buried with tools that might help them in the afterlife, so maybe a domesticated animal was like a tool when it came to being assessed as a useful grave good. They didn't necessarily love that stone axe or clay jar, but they were useful in this world and so, presumably, would also be handy for whatever lay beyond in the next. But it's difficult to argue that a desperately ill puppy would be useful in the same way.

We don't know who died first, or how Bonnie's people died other than it was likely, since there were no signs of trauma, from disease (but not distemper, people don't get that). Is it such a stretch to imagine that when Bonnie was suffering with distemper, some effort was being made to treat her? I'd say not at all; it's logical. Recall that she lived for five to nine weeks after contracting distemper. This is impressive. Even with our sophisticated 21st-century veterinary science, the mortality rate for canine distemper remains high.** That Bonnie survived so long without modern

* The experiment is ongoing. A website which was in operation from 2011 to 2014 purported to make the foxes available for purchase as pets. It's unclear how many, if any, were sold.

** Distemper mortality rates being 50 percent in adult dogs and 80 percent in young puppies.

medical help is striking. She had to have both been very lucky and the recipient of a lot of attention and care.

Keep in mind that our ancestors at that time were as intelligent as we are now but had far fewer distractions. They had the mental bandwidth to really focus on the things that were important to them. Furthermore, researchers believe that during this period, the Upper Paleolithic, most people only had to work four hours a day to provide for their needs. To be sure, they lacked science and literacy, and their view of the world and its workings was infused with what we now recognize as magical thinking and superstition, but they could take all this free time to sort out simple cause and effect when it came to medical treatments. For example, traces of yarrow and camomile have been found on a tooth with a dental abscess in a Neanderthal woman from 50,000 years ago. Neither of these plants are especially tasty or nutritious, but they do have medicinal properties.

Did Bonnie's people also give her some sort of herb like yarrow to try to help her? Did they attempt to apply whatever medical art they had learned for humans, hoping it would help her too? We don't know, but it's reasonable to assume so, and consequently it's reasonable to think of her as the world's first known veterinary patient, so long as we permit generously broad definitions for the words "veterinary" and "patient." My point, however, is that people have loved animals for thousands and thousands of years, and in loving them they have tried to care for them medically. In some parts of the world, including the West, that love was superseded for long periods by a more practical view of animals, but it was always there, waiting to come to the fore again.

From the outset, though, just as with the Napoleonic War horses, there were two primary motives to help sick or injured animals. We can be reasonably sure that because Bonnie was a sick puppy and was buried with the family, if she was being treated it was because she was loved. This is the first motive — the emotional. But if Bonnie had been older and had, say, been a first-rate hunting companion who became ill, then her theoretical treatment would fall into the second category of motivation — the practical. In that case, treating a sick or injured animal was like trying to repair a damaged tool. To be sure, both motivations can be present at the same time, a situation familiar to small-scale farmers everywhere today. You can have genuine affection for your prize dairy cow, but she is also important to your livelihood and has a crisply definable dollar value. These two reasons to help sick animals remain essential features of veterinary medicine to this day. Sometimes they are completely distinct, and even at odds with each other, and sometimes they merge and reinforce each other.

In many places and at many times throughout history, there was another powerful element that motivated humans to provide medical care to animals — they were not seen as truly separate from us. This may well have been another factor in Bonnie's case. For people who held this view, the Circle of Concern was vast and essentially boundless. For better or worse, this element plays no role in modern veterinary medicine or in its history as an organized profession in the West, but it's important to acknowledge. Before the advent of organized religion, and especially that of the monotheistic creeds, most people believed in a form of what we would now refer to as nature worship or animism, where the entire world is imbued with divinity and where the various inhabitants, animate and inanimate, are not seen as necessarily distinct from each other. We have all heard of people referring to their "spirit animal," and while this can sometimes be tongue-in-cheek ("Yeah, I'm sure the sloth is my spirit animal"), it's rooted in a very real belief held in many cultures around the world that there is a spiritual connection between specific people and specific animals. These animals would often also be seen as siblings in a metaphorical sense so powerful that it may as well be literal.

This belief is familiar to many people as a traditional North American Indigenous practice, but pre-Christian Europeans had similar beliefs. Some of these beliefs trickled down through the centuries to appear in modern forms of paganism, such as Wicca. Spirit animals in European paganism are sometimes referred to as "familiars," the most well-known example of which is the iconic witch's black cat. But black cats are simply the only ones whose stories have made it through the centuries to become part of the popular modern imagination. In fact, familiars could be all sorts of animals, including dogs, ferrets, toads, rats, ferrets, hares, wasps, butterflies, pigs, sheep, horses, and a variety of bird species.

In a case during the English Civil War in the 1640s, it was a dog. Prince Rupert, the military leader of the Royalists, was accused of using a familiar.* He often took Boye, his large poodle, into battle with him. The enemy Parliamentarians, who were usually devout Protestants, were terrified of this dog. This is hard to understand, and even a little comical, when you look at Boye's portrait, depicting a friendly looking fluffy white poodle. But the fear came from a widespread rumour that the animal possessed supernatural powers. According to this story, Prince Rupert used Boye's powers to wicked advantage on the battlefield. Boye apparently also protected his master by catching bullets aimed at Prince Rupert in his mouth. He was also reported to have learned to lift his leg to pee whenever the name of the Parliamentarian leader, John Pym, was mentioned. The darkest rumour, however, was that Boye was Satan in disguise (white colour notwithstanding). This is not as strange as it sounds, as at the time all manner of devilry was ascribed to Rupert's Royalist Catholics. Superstition as propaganda. Eventually Boye was shot and killed by a brave Parliamentarian, allegedly using a silver bullet.

We can be confident that Prince Rupert did not view Boye as his familiar, only as his cherished companion. Boye even slept in the prince's bed. But Prince Rupert was probably happy to let the enemy think that some magic was afoot. Nor did Sarah Good or Martha Corey have a special relationship with "a yellow bird" in Salem during the cruel and infamously misogynist witch trials. Viewed from the outside, familiars were always associated with evildoing; but viewed from the inside, from the perspective of those who held pagan beliefs, this was not the case. Pagan familiars were simply considered extensions of the person, and therefore no more or less likely to be used for ill purposes than the person would

* For you Canadiana buffs, this is the same Prince Rupert that Rupertsland was named after. And, of course, a city on the north coast of British Columbia.

be motivated to use any other resource. A familiar could just as easily be a force for good. The point of this digression is that the nature worship common to our distant ancestors often involved a closeness to animals that we as modern people don't understand at the same instinctive level. From a veterinary perspective, it makes perfect sense that people who felt that animals were extensions of themselves would go to great lengths to provide medical care for them.

Shamanism is an even more extreme expression of this intimacy with the natural world. In shamanic traditions throughout North Asia and the Americas, shamans believed that they actually became animals. Through elaborate rituals, they transformed themselves into ravens, or wolverines, or hawks, or bears, or any other animal that would grant them a special power or perspective. I don't hesitate to say that this is, of course, impossible in the rigorously literal sense, but the depth of belief in these transformations speaks again to the importance of animals to people inculcated with this world view. How would you view an injured animal, and your duty towards it, if you knew that it might actually be a person?

Shamanic cultures which have survived into the modern era, such as the Nenets reindeer herders of Siberia, still use traditional medicines to treat their animals, but the utilitarian reasons are difficult to tease apart from any spiritual and emotional connection.

Outside the Abrahamic religions (Judaism, Christianity, Islam), belief in reincarnation is another example in which the relationship between humans and animals is more direct, personal, and intimate, and it provides a clearer illustration of how that relationship led to a form of veterinary care. For example, Hinduism dictates that animals have souls and that these souls will eventually enter human bodies in their journey to become closer to God. This belief still places animals below humans in a hierarchy, but it accords them far more respect than do religions that deny souls to animals. Traditional Indian Ayurvedic medicine was used to treat animals as early as 4000 BCE. The Rigveda Ayurvedic text, written between 2000 and 4000 BCE, describes physicians using its principles to treat both humans and animals. We'll spend some time with an early Ayurvedic practitioner in a later chapter, but to keep to our timeline, we need to move a few thousand kilometres east to Mesopotamia around 2000 BCE.

But before we do so, a last word about Bonnie and "the old woman." It's not an accident that I chose a woman to be the first of the 23 representative animal healers and veterinarians this book is structured around. Not only was the art of healing often the purview of women in traditional hunter-gatherer societies, but also, as you'll eventually see, the art of healing animals has become predominantly female led in modern times. Full circle, perhaps.

CHAPTER THREE

Ubar

Ubar mopped his brow. The midday heat was intense, even in the shade of the date palms beside the canal. He blamed the heat, but he was also sweating because he was nervous. Rimush's ox was very valuable. Ten shekels, Rimush said. If the ox died, Ubar would have to pay this angry and skeptical man two and a half shekels in compensation, and if the ox lived, his fee for the procedure would be one-sixth of a shekel. Ubar was good with sums. He knew that for surgery to be profitable, he must save more than 15 oxen for each one that died. But it was only an abscess, so the chance of killing the ox was low. It made sense to do it. But if the Gods were unkind, and some accident should occur during the procedure, he would suffer not only the loss of money, but also the loss of reputation. Rimush was well connected. There were other cattle doctors in Babylon, only too happy to take Ubar's clients. But if Ubar succeeded, and he probably would, a sixth of a shekel was not bad for an hour's work. A potter or brewer or weaver would have to toil all day to earn that. Although, on the other hand, a surgeon working on people could be paid up to five shekels for what Ubar was about to do for this ox.

Five shekels! Thirty times what Ubar would be paid!* He supposed the responsibility was so much higher, but even so.

These were the thoughts that went through Ubar's mind as he sharpened his knife while Rimush looked on from the shade of the wall with his arms crossed. When the blade was honed to his satisfaction, he muttered a prayer to Shakkan, the God of cattle, before motioning to Rimush's slave to step forward.

"Bind your master's ox tightly to the post," he said.

Rimush's slave did so, wordlessly.

"Tighter." When this was done, Ubar nodded. "Now pull on his nose so that he looks away from me, towards the wall there."

The abscess, a swelling the size of a ripe pomegranate, was in the ox's right foreleg, in the fetlock, just above the hoof. It had made him lame and stopped him from ploughing Rimush's valuable barley fields just beyond the city's walls.

The knife's edge was as keen as it ever had been, and the ox was as ready as he ever would be. Rimush grunted with what Ubar assumed was impatience.

Ubar smiled at Rimush and nodded. Rimush did not smile or nod back. The slave glanced at Ubar, as if expecting further instructions.

There were none. There was no point in delaying.

Ubar mopped his brow one more time, breathed another quick prayer to Shakkan, and plunged his knife into the softest spot on the lower part of the abscess.

Ubar and Rimush may be imaginary, but the scenario is realistic. While India can claim the first clearly documented veterinary hospital, Mesopotamia, Egypt, and China also had ancient animal healing traditions. Around 2100 BCE in Mesopotamia (or 3000 BCE, depending on which source

* In case you're curious, today fees for procedures on humans are only about twice as much as for animals, on average.

you believe), a man named Urlugaledinna was referred to as "an expert in healing animals." We don't know much about him, and he likely worked on humans as well as animals, so we can't call him the "first *veterinarian.*" To minimize confusion, that word should be reserved for doctors who exclusively work to heal animals. Urlugaledinna's personal cylinder seal depicts a pair of tongs likely used for minor surgical procedures.

A few centuries later, around 1750 BCE, the famous Babylonian Code of Hammurabi was written. This is the first legal code in history to come down to us intact. It is a set of 281 laws covering all manner of potential scenarios, crimes, and disputes, including, remarkably, two veterinary decrees:

> 224. If a veterinary surgeon performs a serious operation on an ass or an ox, and cure it, the owner shall pay the surgeon one-sixth of a shekel as a fee.
>
> 225. If he performs a serious operation on an ass or ox, and kills it, he shall pay the owner one-fourth of its value.

A shekel was about ten grams of silver, and a sixth of that may seem like a good wage for what cannot have been more than an hour or two of work, given that for only 20 shekels you could buy a slave (a whole human being!), but as we saw in Ubar's story, the cost of failure was high.

The ancient Babylonians were not only concerned about the health of asses and oxen, but also with that of dogs. Although Hammurabi's Code does not mention them, in the surviving fragments of the Code of Eshnunna, which may actually predate Hammurabi, there is a requirement for the owners of rabid dogs to prevent them from biting people. This seems curiously considerate of the dog's perspective, given that euthanasia of rabid animals was otherwise the norm throughout history, even today. Rabies is uniformly fatal in dogs, and in people for that matter, so it seems likely that the ancient Babylonians had a loose definition of "rabies" that may have included all causes of inexplicable aggression. Even so, it's a remarkably progressive law. Rabies, by whatever definition, seems to have been a subject of interest to the Babylonians, as one of their Goddesses, Gula, was specifically tasked with protecting people against rabies, in addition to her other more general healing duties. Tellingly, her assistant is often depicted to be a dog.

Later, in the sixth century BCE, the remarkably dog-friendly Zoroastrian religion developed next door in Persia.* Dogs were holy animals for the Zoroastrians and were led to the bedside of dying people to ward off evil spirits at this critical time. The Zoroastrian sacred book, the Zend-Avesta, makes numerous references to the treatment of canine diseases and lays out a long series of penalties for dog beaters.

> Zarathushtra asked Ahura Mazda: "O Ahura Mazda, most beneficent Spirit, Maker of the material world, thou Holy One! He who smites one of those water-dogs . . . so that he gives up the ghost and the soul parts from the body, what is the penalty that he shall pay?"
>
> Ahura Mazda answered: "He shall pay ten thousand stripes with the Aspaheastra, ten thousand stripes with the Sraosho-charana. He shall godly and piously bring unto the fire of Ahura Mazda ten thousand loads of hard, well dried, well examined wood, to redeem his own soul."

And that was just the start. The law goes on for 18 stanzas of cumulative penalties, ending in:

> He shall put into repair twice nine stables that are out of repair. He shall cleanse twice nine dogs from . . . all the diseases that are produced on the body of a dog. He shall treat twice nine godly men to their fill of meat, bread, strong drink, and wine.**

* Once with an estimated 40 million adherents, there are now fewer than 200,000 people who practise Zoroastrianism, mostly Parsis from India and Pakistan.

** https://avesta.org/vendidad/vd14sbc.htm

Moreover, and here's the kicker, anyone who had done harm to the animal was barred from heaven, so it was common for dying people to confess their sins to dogs and seek forgiveness. It was handy that they were there anyway to guard against those evil spirits.

At almost the same time in Egypt, we find the Kahun Papyri, a set of documents which list treatments for cattle, dogs, cats, birds, and, believe it or not, fish. It also makes special mention of the importance of washing one's hands before and after treating a sick animal. That ancient Egyptians practised a form of veterinary medicine is not surprising when one considers that they venerated many animals, mummifying countless thousands of cats especially, but also dogs, ibises, snakes, and even crocodiles, to name just a few. At one site alone, in Saqqara, archaeologists have found more than a hundred thousand falcon mummies, as well as four thousand baboons.* Animals were not only viewed as connected to the divine, but also some were treasured as pets. Ancient temple and tomb paintings depict scenes of delight and affection between people and their pet cats, dogs, and monkeys. An interesting aside is that the ancient Egyptians also trained cats to help hunt birds, something not seen again in history. One assumes it was difficult to get the cats to listen consistently.

The Kahun Papyri also discusses trypanosomiasis, or sleeping sickness, which is spread between animals and humans by tsetse flies. This is the first documented reference to a zoonosis (human disease passed from animals), an area of study which would later become a cornerstone of modern veterinary medicine. Some of the treatments are still in use today, such as hot cautery (mind you, under anaesthesia these days), copper compounds, and enemas. The use of honey as an antiseptic on wounds is also mentioned, which has made a return in the last few decades after having disappeared from mainstream practice for millennia. Some specific cases are described as well, such as what was likely necrotic colitis in a bull. The veterinarian chants a prayer, inserts

* Associated with the Gods Horus and Babi, respectively.

his arm into the rectum of the bull (see, some things haven't changed), and then scoops out the infected material while rubbing the bull on the back with his free hand. He does this several times, washing his hands and arms in between. Eventually there is success, and the bull is cured. Remarkable.

In China, the mythical Emperor Fuxi is credited with founding veterinary medicine. According to another legend, veterinary acupuncture was discovered when lame horses were cured of their lameness after being struck by arrows in battle, presumably in specific locations and not too deep. Yet another legend tells the story of the horse doctor Ma Shih-huang. One day, a normally fierce dragon came to him, no longer looking so fierce with his drooping ears and a gaping mouth that he couldn't properly close. Ma Shih-huang examined the dragon and determined that the correct treatment was to make incisions in the lips and to administer a draught of liquorice. Happily, he was correct, and the dragon was cured. After that, more and more dragons came to see him. Ma Shih-huang had become the world's first, and we can safely presume only, dragon doctor. Then one day, according to the legend, the dragons carried him off and he was never seen again. It is unclear whether this is meant to be a happy ending or a sad one.

The non-mythical beginnings of veterinary medicine in China are uncertain, but we know that by around 2000 BCE both acupuncture and herbal remedies were being used in animals. This first individual recorded so doing was an unnamed coach driver for King Mu of the Zhou Dynasty. This coach driver apparently had a talent for curing horses of summer fever by bleeding them. I suspect that he mostly had a talent for being lucky and timing his bleeding when the fever was about to break anyway. Then, around 650 BCE a General Sun Yang, also known as Bai Le, wrote the *Bai Le Zhen Jing* (*Bai Le's Canon of Veterinary Medicine*), the first Chinese veterinary book. Bai Le is considered the father of veterinary acupuncture, and sometimes generally of veterinary medicine in China, although there are other contenders for that title. Bai Le's text focused on horses, but later works included references to pigs, camels, cattle, chickens, geese, ducks, goats, and sheep.

Veterinary medicine was considered prestigious in China. "Service of Physicians for Animals" was one of the five official health services in the Zhou Dynasty, around 400 BCE. Veterinarians in this service had specific protocols for examining horses. They sprinkled their patients with medicinal herbs to comfort them, walked them carefully to observe their gait, and paid special attention to the pulse. The government tallied each vet's successes and failures and paid them based on the net success rate.

Meanwhile, around the same time as Bai Le, across the Himalayas in India, a mysterious sage named Palakapya wrote the Hastyayurveda, a text describing 315 elephant diseases and their Ayurvedic treatments. He is mysterious because he is also described as being the contemporary of a mythical king and as being the incarnation of Dhanvantari, the mythical physician to the Gods and founder of Ayurveda. One legend has him being found as an orphan in the woods. Elephants picked up the baby Palakapya and delivered him to the doors of a hermitage. Another legend details that Palakapya was the son of the sage Samagayana and a female elephant, who had somehow managed to drink Samagayana's urine with semen mixed in it. No, let's not explore this. In any case, the book is real, but the man's existence, uncertain.

CHAPTER FOUR

Palakapya

"Sri Palakapya," Bharav said. He addressed an old man lying on a bamboo platform under a thatched sunshade. The old man's eyes were closed. It wasn't clear whether he was asleep or just resting. He didn't stir.

Bharav, a young soldier in armour of overlapping leather scales, took off his helmet and glanced around, hoping to spot someone who could help, someone with more authority to wake the great man. But there was no one. Just a small black dog, also asleep or resting.

He took a step forward. "Sri Palakapya," he said, his voice a little louder this time. Somewhere above them, a monkey screamed.

Still nothing.

Bharav pursed his lips and considered the situation. The sage was known for his temper, but then so was General Kunaal, who had sent him. Bharav finally decided that it was better to anger a thin-limbed old man than a powerful soldier, so he took another step forward, "Sri Palakapya!"

The dog stirred and looked up at him. The man didn't respond at first, but then he smiled and, after a few long seconds, opened his eyes.

He rolled over and propped himself up on one elbow. "Namaste, child. I was having a beautiful dream, and I wanted it to finish before waking. I apologize."

Not angry, thank goodness.

"Namaste, Sri Palakapya. No, I apologize for disturbing you." Bharav looked at the ground as he said this.

Palakapya coughed loudly and then laughed. "Now that we are finished apologizing to each other, please say what brings you to me."

"General Kunaal sent me. One of his elephants is unwell. It is an important elephant, a big bull named Manvir."

"Ah yes, I know Manvir. A handsome beast. Fearsome in battle. In what way is he unwell?"

"He is restless. Stomping, shifting, and trumpeting. He will not settle."

"Mmm, I heard him this morning in the distance. I wondered." Palakapya sat up and groaned lightly. "What has he been eating?" He was already forming a suspicion.

"Bananas, Sri. Many bananas. He broke into the house where they are stored."

"As I thought. Run on ahead, as I will be slow coming. Tell the general that we must do basti. Tell him to open my book to the fifth chapter of the 'Uttarasthana' section. This chapter is called 'Vastidanakathanam.' It has all the necessary instructions. He can begin the preparations while Chiti and I make our way there." At the mention of her name, the little dog stood up and began wagging her tail.

"'Uttarasthana' section. Chapter five, 'Vastidanakathanam.' Prepare for basti," Bharav said.

"Excellent, thank you. You may go now." Palakapya smiled and gestured in the direction of the elephant compound.

Bharav nodded and turned to go, but then turned back. "Forgive me, Sri, but basti? In a bull elephant?"

Palakapya just laughed and waved him away.

Bharav jogged along the narrow path through the trees and tried to picture what was about to happen. He was unable to conjure a mental image that wasn't both absurd and terrifying.

The elephant compound looked chaotic when Palakapya and Chiti arrived an hour later. People were moving in all directions. There was a lot of shouting. And above it all, the loud trumpeting Manvir, who stood in the middle of all the activity, like the large grey eye of a storm.

An obese man in rhinoceros hide armour spotted Palakapya and motioned to him. Even if Palakapya didn't know him already, it would have been obvious that it was General Kunaal because of his elaborate golden helmet and the ring of attendants looking nervously at him. The general's face was bright red. He was angry or hot, or likely both.

"Namaste, Sri Kunaal."

"Namaste, Sri Palakapya. Thank you for coming. Manvir is very valuable. He is my best warrior elephant and should sire many more like him. I cannot lose him. We have prepared as you have asked . . . but, Sage, basti for an elephant? Is that not —"

"Dangerous? Perhaps, if performed incorrectly. But in the tenfold times I did so for your predecessor's elephants, we did not lose a single mahout or handler." Palakapya chuckled. "But a few had surprise dung baths! Quite healthy, actually, though perhaps a little unpleasant."

The general wrinkled his nose and bowed his head slightly. "You are the doctor, Sri. Would you like to inspect the tube? It is precisely 68 angulas[*] long, as you have written."

"Yes, please. Bamboo? Or wood? Or copper?"

"Not copper on such short notice, so bamboo. I hope that is as effective."

"It is. I prefer copper for washing infected tissue, but that is not the case here," Palakapya said.

The general nodded to one of his attendants, who then scurried off.

"And the solution, General?" Palakapya said. "Indrayava, Kustha, Madhuka? Dasamula? Asvattha? Vata? Asvakarna? Chitraka? You have all of those?"[**]

"Yes, Sri, and the oil, rock salt, and jaggery sugar as well, of course. Sri Daivik uses many of these when he is treating my soldiers," the general said.

---

* An angula is the length of the middle phalanx of the index finger, so 68 angulas is roughly two metres, or six feet.

** These are all trees, herbs, and roots that are still in use as gastrointestinal remedies.

"And your bhahuyantra is strong enough for Manvir?"

"My workers have reinforced it. They have doubled the leather straps and have added more wood to the places where Manvir is strongest, at his hips and shoulders." The general pointed across the compound to a what looked like a complex wooden cage, about the size and shape of an elephant. A couple of workers appeared to be putting finishing touches on it. Palakapya preferred not to use a bhahuyantra like this to restrain his elephant patients, but with a bull like Manvir it was unfortunately necessary. Especially since he intended to insert a long bamboo tube into Manvir's anus.

Everything appeared to be in readiness 20 minutes later. A half-dozen priests had been rounded up to chant soothing songs to Manvir, and the elephant's best mahout, a short bow-legged man named Gaurang, was summoned to persuade him to lay down inside the bhahuyantra apparatus, which had been opened like a clamshell.

This was no mean feat. The elephant was agitated. He was clearly in discomfort from his constipation, and fearful of what the humans intended to do with him. But after a few minutes, whether it was due to the chanting, or the years of training, or Gaurang's skill, Manvir slowly lowered himself to the ground and rolled onto his side and into the bhahuyantra.

The chanting continued as nervous-looking workers closed the bhahuyantra around Manvir and secured it with the leather straps. Gaurang talked to Manvir quietly through all of this, reassuring him, but also made it clear that he was in charge.

Meanwhile, Palakapya adjusted the oil and herb enema mixture until it was to his liking. They had used too little Vata, but that was easily remedied. And it wasn't warm enough, so a coal brazier was brought. Body temperature was ideal. He glanced over at the unfortunate elephant and his mahout from time to time, assessing the optimal time to begin. It was important to be patient and wait until the elephant was as calm as he could be under the circumstances. But it was equally important not to wait too long, as the period of maximum calmness could be brief.

Gaurang caught his eye and gave him a sharp nod.

Manvir was as ready as he was going to be.

Palakapya could have delegated the insertion of the tube and attachment of the bladder filled with the basti solution to a helper. He had done so before, but always with lesser beasts. With Manvir, all the technique and finesse the old man commanded would be critical. Decades of experience had taught him how quickly to insert the tube, when to pause, when to twist, when to back off a little, and most critically of all, when to squeeze the bladder and how forcefully to send the warm herbal oil coursing into the depths of the patient.

The sage nodded back to the mahout and motioned to the priests to intensify the chant. He oiled the tip of the tube and approached his patient. The insertion was a simple matter, but it still elicited gasps from the onlookers. It always did. Manvir was unmoved. Palakapya was unsurprised. He knew that the elephant would be preoccupied with the discomfort of his constipation, and with the indignity of his confinement.

Minute tremors transmitted down the tube informed Palakapya's fingertips. Despite Manvir's great size, only 62 angulas' length of the tube was needed. This was the spot. With the help of Bharav, who was visibly trembling, and another even more nervous-looking soldier, Palakapya manoeuvred the enormous leather bladder onto the end of the tube and secured it in place. He muttered a prayer to himself and then began to squeeze the bladder by rolling it up from the far end.

Manvir twitched. There was a soft, almost imperceptible, rumbling sound from deep inside of him.

Palakapya stepped aside.

In addition to detailed instructions on how to administer enemas, the Hastyayurveda carefully describes how to examine an elephant and determine whether it is healthy. Palakapya gives special attention to the sounds an elephant makes. It's a good sign if its voice is like that of thunder, a conch, a swan, a tiger, or a bull, but if it is like a parrot, cow, monkey, or boar, something is wrong. Among those 315 elephant diseases described, the most prominent were cracked feet, pneumonia, inflamed eyes, two kinds of tuberculosis, dental abscesses, and, as our little sketch illustrates, digestive disturbances, including constipation.

The dental abscesses are interesting too, though. Imagine a tusk becoming abscessed. Dental abscesses are normally very painful, so one in a tusk must cause the poor animal agony given the size of this tooth and the fact that it protrudes from the mouth and is easily accidentally bumped. The bhahuyantra described above was frequently used for restraint, which may seem cruel to 21st-century readers, but it was really no different in spirit than the restraints used to keep human surgical patients still during painful or unpleasant procedures at the time.* Palakapya gives every indication of having genuine concern for the welfare of the elephants, although, unfortunately, he also recommended giving them alcohol to make them more aggressive in battle.

However, too little is known about the real Palakapya, if he even existed, to award him with the title "world's first veterinarian." Instead, that honour must be granted to another Indian, Shalihotra, who was the first physician, documented anywhere, who devoted himself solely to the care of animals. Around 300 BCE he wrote the *Shalihotra Samhita* (Shalihotra's Encyclopedia), which detailed anatomy, physiology, surgery, and medicine, including preventative medicine, for both horses and elephants. His name was so synonymous with veterinary medicine that the earliest word for veterinarian, "salihotriya," is derived from it. Based on his work, approximately 50 years later the great Indian King Ashoka opened the first veterinary hospital in the world.

Often considered one of India's greatest emperors, Ashoka fought in many battles where horses and elephants were crucial to his famous victories. The last of these wars, against Kalinga (modern Odisha in eastern India), resulted in more than a hundred thousand deaths. Something shifted inside the hardened warrior-emperor when he witnessed this mass slaughter, and in 260 BCE he converted to Buddhism and vowed to stop waging war. Nonetheless, any general worth his gold stars will tell you

* For insight into the arrestingly gruesome world of surgery before the invention of anaesthesia, please have a look at Ira Rutkow's *Empire of the Scalpel.*

that when you live in a violent part of the world in violent times, your best guarantee to remain at peace is to be well defended. Horses and elephants were a key part of that defence. It's reasonable to believe that having a good defence was an important motivation in opening this hospital. It's also worth noting that Ashokan India was a state which believed itself to be the most modern and forward-looking in the world, so it would have felt natural and right to innovate and be the first in many fields, including veterinary medicine. Unfortunately, the location of Ashoka's veterinary hospital is unknown, although it is reasonable to guess it to have been in his capital city, Pataliputra (beside modern Patna). It is well known that Ashoka loved his horse, Pawan, so if he was going to decree the establishment of a veterinary hospital, why not have it nearby?

As is so often the case in the story of veterinary medicine, the practical reasons for attempting to heal injured and ill animals (I need this horse for my army) would have been mingled with the emotional (I love this horse) and possibly the spiritual (this horse could be my future grandson).

Meanwhile, to the west, in the Mediterranean, another veterinary tradition was developing in fits and starts, culminating in the fifth century CE with a Roman military reformer named Vegetius.

CHAPTER FIVE

Vegetius

Rome was a nightmare. It was a nightmare for men, other than those in the very highest echelons, and it was a nightmare for horses. Vegetius considered this inarguable fact as he surveyed the motley array of lame horses in his courtyard. Most of them had become lame because of the state of the streets. Constantinople was better. A more modern city with wider, well-maintained streets, and far less garbage and rubble laying about. He could relocate there, but this was where he was needed. This was where his work was. At least for now. He had been all over the empire, so who knew where he might need to go next? So long as it was soon.

"Walk him again, this time at a trot. Then turn tightly at the wall and trot back," he said.

The slave did as he was told while Vegetius, the famous Publius Flavius Vegetius Renatus, stood in the shade and watched, hand cupping his chin, eyes narrowed.

Clip-clop, clip-clop, clip-clop.

Any lameness was subtle. Too subtle for Vegetius, especially as it had already been a long day. It was time to rest and drink wine. Where was Flavian Triarius anyway? They hadn't seen each other in over year as Triarius had been away in Greece. Vegetius had been looking forward to the reunion for weeks. But his friend, as usual, was late.

The paving stones in this courtyard were smooth and even, unlike in the streets beyond the gate. It should make a lameness easier to discern without having to factor in the poor animal needing to manoeuvre around holes and jagged piles of broken pavers. Picturing the streets, Vegetius considered, not for the first time, that even Ravenna would be better. He could move back there. It wasn't nearly as far as Constantinople. He could work on his new project with more peace and solitude. Not complete peace and solitude, mind you. And there were rumours that the western capital would be transferred there soon because of its better situation. That would put an end to the peace . . .

Vegetius snapped out of his reverie. The slave held the horse, a beautiful dappled grey mare, by the bridle and looked at him expectantly.

"The owner says he is lame?" Vegetius asked. He made no effort to hide his irritation.

"Yes, sir. Front left leg, he said." The slave looked at the ground as he said this, but his voice was clear and confident.

"Well, that may be so, but he's of no use to me. Send him away and bring me a truly lame one."

"This is Senator Arminius Lepidus's son's horse, sir."

Vegetius sighed. "All right. Tell him to have his best groom bleed the animal from the neck and mix the blood with vinegar. It must be a very sharp vinegar, and it must be mixed thoroughly. Equal parts." He paused. "Can you remember all this?"

"Yes, sir." The slave glanced up at Vegetius and patted the horse on the neck.

"Good. This must then be anointed to the fetlock of the left front leg. The horse likely has a mild articular distemper. I cannot be sure that it is in the fetlock rather than the knee, but we will do no harm to begin treating there. It will draw out any morbid corruptions that may be present. The treatment of the fetlock is much simpler than the treatment of the knee. Should the senator's son find that his horse is still lame, he may need to take the trouble to treat the knee."

"What is that treatment, sir?"

Vegetius looked at the man. He couldn't remember this slave's name. He was new. But he seemed bright enough. "You are certain you can remember it with fidelity?"

"I am certain. In my previous life, I was responsible for remembering everything about my tribe's cattle. We did not have writing."

"Germania?"

"Yes, sir. Marcomanni."

Vegetius smiled and nodded. "The great Marcus Aurelius fought both with and against the Marcomanni two hundred years ago."

"In three wars."

"You know your history well. Then perhaps you will be able to remember this recipe."

The slave nodded. He was no longer looking at the ground, but he didn't meet Vegetius's piercing eyes either. He looked at a spot somewhere over the older man's left shoulder while continuing to pat the horse.

"If the blood and sharp vinegar on the fetlock do not make the horse sound, then they are to treat the knee. For this they will need blood and vinegar again, three sextarii* of the latter, and then mix that with a pound each of white clay, nettles, cumin, resin, and tar. To this add a handful of salt and of ox dung."

The slave nodded and repeated what Vegetius had said.

"Well done."

"Sir, you did not mention how much blood."

"Excellent question! Just enough to make a paste that allows itself to be smeared evenly over the knee, without being too thin or too thick. And I should mention that the blood needs to be drawn from the shoulder of the same leg."

---

* One sextarius was 546 millilitres, or just over a pint.

"Yes, sir. Thank you."

Just then a tall slender man entered the courtyard. "Greetings, Flavian Triarius! Come sit here with me!" Not only was the shade under the eaves pleasant at this time of the afternoon, but also from here Vegetius could describe his project to his friend with the horses visible in their stalls off to the side as a suitable backdrop.

"So, old friend," Triarius said after he had made himself comfortable and accepted a glass of wine from the serving girl. "Let us drink to the emperor's health and, more importantly, to our own. And then you can tell me why the great Publius Flavius Vegetius Renatus, author of the renowned *Epitoma rei militaris*, is wiling his time with these decrepit nags. Although, mind you, that last beast was a fine specimen. Sound-looking to my inexpert eye."

Vegetius didn't reply. Triarius wasn't the first to ask this question. He smiled weakly at his friend and raised his glass.

"Therefore, let him who desires peace prepare for war!" Triarius took a deep swallow from his wine. "That was what you wrote, yet here you are, preparing for a ride in the country? Do you not desire peace?" He laughed. "This is an excellent wine, by the way. Apulian?"

"Illyrian."

"Interesting, but in all earnestness, my friend, why is the greatest military strategist of our age spending his days gazing at horse's feet?"

Vegetius took a long sip and considered his answer. Finally, he said, "I probably should have written, 'Therefore, let him who desires peace prepare to have his words ignored.'"

"That cannot be true." Triarius said this with his mouth half full of roasted pigeon breast. He was known for an appetite that belied his thin frame.

"But although it cannot be, it is," Vegetius said. "There were many fine words of praise, all the way up to Emperor Theodosius, who commended me on my analysis. But it is clear now that these words will not become actions. I, better than anyone, understand that the Roman army of two hundred years ago cannot be recreated, but I had hoped that lessons could be drawn and reforms implemented."

"The best of the past blended with the best of the present," Triarius said, and nodded.

"Precisely, although the best of the past is better than the best of the present. Decline does seem to be the way of the world."

Triarius held his glass out to the side. The serving girl, who had been standing silently in a deep shadow beside a column, stepped forward and filled it. "I suppose it does seem that way," he said. "Except with horses? Is that why you have turned your attention there?"

"As always, you are very perceptive, my friend," Vegetius said. "I decided to turn from a general overview of military planning and strategy to a detailed look at one specific aspect: the horse. They have not declined but are as they always have been. The army and the empire and society may be losing their way, but not the horse."

"To the horse!" Triarius raised his glass.

"To the horse," Vegetius replied. "And the cavalryman of today cares for his horse as much as he did in Marcus Aurelius's time. The barbarians pressing on our frontiers are often mounted on horseback, so cavalry is more important to Rome than it ever was."

"Agreed."

"Growing up on a farm and working there with the horse healers, I learned a lot. I have also read the Greeks and our Roman forebears on the subject. I wager that I have read more than anyone in the empire on the diseases of horses and their cures. But everything before was poorly written."

"You are a good writer, my friend." Triarius reached for a pork hock. "Famed across the empire for it, I dare say."

"Thank you." Vegetius paused to take a sip of wine. "This *Digesta artis mulomedicinae* I am writing will pull together all that is known. Perhaps I will succeed in helping the army in this one specific subject when I approach it in depth, rather than the perhaps overambitious breadth of *Epitoma rei militaris*."

"Perhaps you are right. But I still think you underrate the impact of *Epitoma*."

Vegetius waved off the remark as if shooing a fly from the meat platter.

Triarius went on, "And evidently, you are not only editing and summarizing what others have written before. You also appear to be doing your own research." He inclined his head towards the stalls across the courtyard.

"I am confirming a few things with my own eye. And word of my interest has gotten around, so people are sending me their animals. For now, this is helpful to my project as I will dissect some of those who are too ill or lame to be saved. To understand how the tissues are arranged is very helpful. But soon I will go somewhere quieter. Writing is difficult here."

"Everything is difficult here!" Triarius exclaimed. "Even if Alaric hadn't sacked it, Rome would be a —"

"Cesspool?"

"Yes, a riotous, calamitous, odiferous, obnoxious cesspool!"

"But with wine from Illyria, serving girls from Circassia, slaves from Germania, and, most importantly, the best libraries in the empire."

"Still?"

"Yes, still. And I need them for now."

Just then the German slave came running back into the courtyard with the dappled grey mare.

"Sir, pardon my interruption. She is showing her lameness now. It began when I went to exercise her outside. Watch, please."

Sure enough, there was a marked offbeat gait favouring the left foreleg.

"It is the fetlock. Thank you." He turned to Triarius and laughed. "The senator's son will be pleased by the simpler cure. The stench of blood, vinegar, nettles, cumin, resin, tar, and ox dung mixed together is something to behold — even for someone used to Rome!"

Vegetius lived in the fifth century CE. The title of his *Digesta artis mulomedicinae* translates to *Guide to Mule Medicine*. Despite the title, the book focuses on horses, so you would think it would have been called *Digesta artis equumedicinae*, but the Romans referred to all people who treated any animals as "mule doctors." Mules were exceptionally important to the army as draft animals, arguably more important than the cavalry horses. Either way, the *Digesta* has no fewer than 240 chapters covering a breathtaking range of topics, including anatomy, breeding, various fevers, and other illnesses. The only English translation dates from 1748 and gives the chapter titles a charmingly archaic ring. Examples include "Of the Cure of Elephantiasis," "Of the General Causes and Cures of the Maul" (no idea what that is), "Of a Physical and Anniversary Remedy," "Stallions Must not Be Emptied by Bleeding," "Of Horses etc. Incommoded by the Stone," "With What Care Worms and Botts May Be Taken Away with the Hand," "Of Horses Affected with Fainting Fits," "Of a Disordered Brain" and, curiously, also "Of Madness in the Head," and finally, my favourite, "Of a Horse that Is Blasted or Planet-Struck."

I presume that "planet-struck" is the 18th-century English translation of a Latin expression for falling afoul of some astrological malevolence. The Romans, for all their cold engineering rationality, were not above indulging in what today we would classify as superstition. As I have likely now aroused your curiosity, this is what Vegetius says (in an 18th-century English gentleman's voice) about the condition:

> If a Horse, or a Mule, etc. be blasted, or Planet-struck, is known by these symptoms; his Lips and Cheek-bones, as also his Nostrils are in part so corrupted and spoiled that he can scarcely bruise, or chew his Food with his Teeth; you shall also find them full of Humours, and when he has his mind to drink, he will plunge his Mouth into the Water up to his Nostrils, because his lips, wherewith his Draught of Water is attracted, are weak.

He goes on to describe what appears to be his all-purpose treatment: blood and sharp vinegar. However, "if the Jaw-bone be Planet-struck, and also stands awry," he recommends adding ox dung.

Speaking of oxen, Vegetius also wrote extensively on cattle medicine, likely because of the importance of the beef supply to the army; he also threw his hat in the rabies ring. As we'll see, most important figures in the history of veterinary medicine had a theory about rabies. And until 1885, all of them were wrong. As a cure, Vegetius suggested boiling the mad dog's liver and feeding it to the person or animal bitten by that dog. On a more positive note, he was also known for making impassioned pleas that the veterinary profession be given more prestige and better pay. Some things never change.

Like Palakapya, Ubar, the old woman in the paleolithic period, and many of the early animal healers we'll meet in the coming chapters, it is unlikely that Vegetius innovated any treatments. Medicine was a body of knowledge arising from, and bounded by, tradition. There was little sense that progress was possible, or even desirable. One's forefathers, and the Gods, knew best. Veterinary practice was more or less static throughout the world until the 18th century. However, a 2012 review of the *Digesta* found Vegetius's grasp of anatomy to be quite accurate, so I added the bit about him performing dissections. But we don't know that he did anything like that. He may have just collated pre-existing knowledge, cleaned it up, and written it in a manner digestible for the times.

Vegetius relied on a number of earlier Greek and Roman sources whom we should take a brief look at before we leave the ancient world. Being mythological, Chiron isn't a "source" per se, but he's worth a mention. This centaur (half man, half horse) was said to have raised Asclepius, the orphaned son of Apollo and a mortal. He taught Asclepius the art of medicine. Asclepius thus became the God of medicine, and his famous staff with a snake winding around it became one of the most recognizable symbols of the profession. While the human-healing profession sometimes tries to lay exclusive claim to him, Chiron, for obvious deeply personal reasons, schooled Asclepius in the healing of animals as well. Ancient coins have been found depicting Asclepius treating cattle. One story in particular tells of Asclepius healing the foot of a rooster. This is interesting because traditionally people sacrificed roosters to him in thanks for recovery from an illness. Did Asclepius then revive them in the afterlife and heal them?

A couple of centuries after Bai Le and Palakapya, Hippocrates (460–370 BCE), the Greek "father of human medicine," was the first European to write about animal health. He is best known today for his eponymous oath. He promoted good health for all species through exercise, fresh air, baths, and relaxation. Importantly, he viewed medicine as a discipline unto itself, not just an adjunct of religion. He stated that diseases were the product of environmental and lifestyle factors, not divine punishment, as was then otherwise widely believed. He also wrote about the kinship between humans and animals, drawing comparisons in the anatomy through dissection. For example, in his *On the Sacred Disease*, he writes, "The brain of man, as in all other animals, is double, and a thin membrane divides it through the middle."*

All of that warms the heart. The great man recognized the commonality of beasts and men almost two-and-a-half millennia ago. And he had all manner of sensible ideas. But Hippocrates is also responsible for a powerful idea that blinded Western medicine for most of those two-and-a-half millennia: his humoral theory. He postulated that humans and animals had four "humours": phlegm (considered to be cold and wet), blood (hot and wet), yellow bile (hot and dry), and black bile (cold and dry). Most diseases were explained by imbalances in the humours, and therefore most remedies were aimed at restoring that balance. The ubiquitous bleeding of patients up until not all that long ago was, for example, meant to remove an excess of the blood humour in patients thought to be suffering from an excess of wet heat, such as with a fever.

The equally famous Aristotle (384–322 BCE) was also a busy dissector, having diligently examined the inner workings of more than 50 different animal species and reported on many diseases in them. He even developed opinions on bee health, as honey production was very important to the ancient Greeks. He got many things wrong, but many right as well. For example, when sheep were dying in Sicily, he correctly determined that it was kidney failure due to overeating and recommended that the sheep be kept off rich pasture until the evening. But he also believed that because he couldn't find any gonads in eels when he dissected them, eels must

* For the full translation of Hippocrates's work, visit http://academics.wellesley.edu/ClassicalStudies/CLCV102/hippocrates.html.

not reproduce. He stated that instead they must arise spontaneously out of the mud. Poof — eels. This is an odd idea for an intelligent man. Isn't a more obvious explanation that the gonads are just too small, indistinct, or well hidden? I suppose his pride couldn't allow him to admit a failure in dissection. Aristotle's bizarre eel theory held sway until well into the Renaissance. To be fair, it took the young Sigmund Freud 400 tries before he found the testes in an eel. Yes, *that* Freud — he studied zoology before he made his indelible mark as a psychoanalyst. This is one of his lesser-known achievements. And an impressive one. Just imagine his state of mind at the 399th eel dissection: "I am not giving up! Ze testes must be in zere somewhere!"

Nonetheless, on the whole, veterinary medicine advanced in ancient Greece to a standard probably comparable to China's and India's at the time. The same cannot be said for ancient Rome.

Cato (234–139 BCE) was the first Roman to write on animal health, with his *De agri cultura* (no translation needed). He grudgingly borrowed some ideas from the Greeks, but also went wildly off on his own with, for example, recommending trying to force a sick bull to swallow a raw egg whole. Cato's snakebite remedy for cattle was curious too — mix fennel with old wine and pour it into the cow's nostrils. And he was also a big fan of putting pig dung on wounds.

Varro (116–27 BCE) had somewhat more useful ideas and in particular can be praised for advising that sick animals be separated from the herd. This may seem blatantly obvious now, but it's worth noting that right up until the 19th century, most respected experts still believed that diseases that affected multiple people or animals came from bad air, so called miasma, rather than spreading from one individual to another. Varro did not understand germ theory, but his reliance on practical observations over theory put him 18 centuries ahead of the curve. He cannot properly be called a veterinarian, though, as this was just one topic he wrote on among a huge number from grammar to geometry to astronomy. He only began writing about agriculture at the age of 79 when he decided that he needed to give his wife advice on how to run their large estates. This sounds like the set-up for a sitcom.

Then there was a step backwards again with Pliny the Elder (23–79 CE). He was another polymath, mostly interested in history, military affairs, and rhetoric, but he took the time to recommend that rabid dogs be treated by removing "the worm beneath their tongue." I hope it comes as no surprise to you that there is no worm beneath the tongue of any dog, rabid or otherwise. What he had observed was part of the lingual frenulum, the band of connective tissue we all have in that location. Putting your fingers in the mouth of a rabid dog to remove this seems ill-advised, not to mention exceedingly difficult from a practical perspective in a time without anaesthesia. Pliny was clearly a theoretician, not a practitioner. I suspect that this also extends to his advice that sick people should drink hot donkey urine. Either that, or he had a wicked sense of humour. And like Cato, he was an enthusiast for the healing properties of pig dung. Why Vegetius favoured ox dung over pig dung is unclear.

One step forward, one step back. Happily, with Pliny's contemporary, Columella (4–70 CE), we see another step forward. His book, *De re rustica* (*On Husbandry*), contains useful advice on dealing with parasites (real ones, not bands of tissue under the tongue), contagious diseases, and breeding livestock. Vegetius made particular use of this book. Of special note is Columella's recommended castration technique for bulls: make a ring from the fibrous part of fennel, place it around the base of the scrotum and then gradually tighten it. This is not much different than the "elastrators" still used today, and possibly even superior as fennel has some antiseptic properties. Alas, Columella also spent a fair bit of time recommending bleeding, but then everyone did. And just like the miasma theory, this particular nonsense persisted right up to the 19th century. More on bleeding later.

Now we can jump ahead three hundred years to around 350 CE in the eastern half of the empire where we have another Chiron (this time a man, not a centaur), who wrote ten books on veterinary medicine. Ten! As you might expect, they're a mixed bag. On the one hand, he recommends a complex nine-step process including wild cucumber purges, mustard poultices, and hot iron cautery for all diseases involving the head, whether rabies, lumps, seizures, fainting, walking in circles . . . you name it. In modern times we refer to this as the shotgun approach — when you don't know what to do, try everything. Although today we also rigorously avoid doing harm. This seems not to have been a priority then. That being said, Chiron's advice for dealing with the prolapse of a mare's uterus is remarkably sound — clean it, lubricate it with oil, push it back in, and inflate a bladder (harvested from an unsuspecting small animal) in the opening to seal it and keep the uterus in place. Twelve days later, deflate the bladder and pull it out. Voila. Chiron was active just before Vegetius and appears to have influenced him substantially.

Before we leave the Romans, I should mention that while their medical heritage is questionable, they did gift us the profession's name. It appears to derive from the Latin veterinarius, meaning "of or having to do with beasts of burden," and veterinarium, which was the compound that housed pack animals beside military forts. Interestingly, however, the Romans themselves, including Vegetius, did not use the word "veterinarius," or any form of it, to refer specifically to doctors working with animals. The closest we come is in the title of Pelagonius's (fourth century CE) book *Ars veterinaria* (*Veterinary Arts*). He was another enthusiastic bleeder, focusing on unwell chariot racing horses. It was the accomplished English scholar and writer Thomas Browne who, in 1646, first used the word "veterinarian" in its modern context, replacing the title "dog-leech," which had been in common use prior to that. Although Browne is unhappily also known for providing testimony at a witch trial condemning two innocent women (you may recall, this is also the time of Prince Rupert and his alleged familiar, Boye), I cannot help but feel grateful to him for his lexical innovation.

With the end of the Roman Empire, Vegetius and all his Greek and Roman forebears were largely forgotten in Europe. The medieval European church forbade the dissection of animals or the study of their diseases.

Veterinary medicine slumbered fitfully in Europe for close to a millennium. Fortunately, across the Mediterranean, the Arabs picked up where the Greeks and Romans left off. I'll weave their story into the unhappier tale of medieval European animal healing. However, before we do that let's take a moment to glance back over our shoulders at the Ancients.

At the beginning of this book, I stated that there were two primary drivers for the medical care of animals: empathy and practicality. In prehistoric times these drivers were inextricably bound together. Right up to recent times, this remained the case for most people with domestic animals living outside of what we are pleased to call civilization. Among the early civilizations, Egypt, India, and the Zoroastrians retained some non-practical motivations for treating animals, whereas, as far as we can tell looking back from all these millennia later, Mesopotamia, China, Greece, and Rome were primarily motivated by the need to keep animals healthy for agriculture and war. It is difficult, however, to imagine that individual animal healers in every place and time did not on occasion feel something for their patients approaching pity or affection. It would only be human to do so.

CHAPTER SIX

Oswald

"I am told you can heal the beasts because you are learned in the Leechbook," the priest said and rubbed his hands together.

Whether he rubbed them together in nervousness or because they were cold, Oswald couldn't tell. Churches were always cold, but St. Martin's was always especially cold. "Yes, I can, Father Wilfred. By the grace of God, I have seen the Leechbook and have committed it to memory. It is a great thing that Master Bald did to cause his scribe Cild to compile it. And a great thing that King Alfred allows men such as me to study it at Winchester."

"Praise be to God," the priest intoned.

"Praise be to God," Oswald repeated, unable to stop his eyes from scanning the priest for any clues as to why he had called for him. Oswald waited for Wilfred to speak, but the priest was silent, still rubbing his hands. He was a small old bald man in a dirty brown cassock. Not impressive to look at, but apparently respected and, at times, even beloved by his parishioners.

"It is my horse, Oswald," the priest said. "He has been elf-shot, and none of my prayers has helped. I even obtained a small phial of oil blessed by St. Cuthbert's holy relics and applied it to his head, but it has not helped either."

"Elf-shot, are you sure?" Oswald asked. These evil sprites were a plague upon the land, causing all manner of mischief and illness, but that they would attack a priest's animal, on sanctified land, shocked him.

"I am sure. Yes, I am sure," the small man said. Oswald noted impatience at the question mixed with sadness and resignation in his voice. "Come." The priest stood.

Oswald followed him outside, through a graveyard of moss-covered crosses standing helter-skelter, to a small, stonewalled paddock around the far side of the church, where an elderly white horse stood grazing. The iron-coloured sky hung low, brushing the tops of distant hills. It would rain again soon.

"See, look at his belly." Wilfred pointed a shaking finger at the horse.

The priest was right. The belly was bloated in a way that could only mean one thing.

Elves.

Elves could be responsible for many diseases, and sometimes it was difficult to tell, but this was their signature ailment. It was good that Wilfred had thought to call for him. Bald's Leechbook was clear on the remedy, and although Oswald had many healing talents for men and beasts, he considered himself something of an expert in dealing with elf-shot horses. He even had the correct tool in his leather satchel, among the various herbs with which he otherwise plied his trade. Mere village priest's prayers and dribbles of holy oil were not enough against this evil.

"I see, Father Wilfred. The poor beast. What a curse this is; but by the grace of God, I know what to do."

The priest nodded and watched wide-eyed as Oswald pulled out an odd-looking knife.

"It is specially made for elf-shot beasts. The haft is made from the horn of a cow who is without calf, and it is studded with three brass nails."

"Yes, I see. How wonderous." The priest crossed himself. "With this you will bleed him? Not with leeches?" He peered at Oswald's open satchel as if expecting to see leeches slithering out.

"No, not bleed, only prick the left ear. But before then I will write Christ's name on the horse's forehead and on each of his limbs. For this I will use the tip of the knife, but I will not scratch him. Christ's name will only be visible to the elf, who will flee immediately. And then I shall strike the horse on the back one time, and he will be made well."

"Thank you, Oswald, I am most grateful. The horse is old, but so am I, and we have been together a long time. When the Lord wishes to take either of us, I will be ready, but this devilry of elves is not the Lord's doing."

"Indeed, it is not." The taller man drew himself up to his full height and squared his broad shoulders. "Now if you can find me a stout stick to strike him with at the end of the treatment, then I can begin."

Bald's Leechbook, written in the ninth century in Anglo-Saxon England, is one of the very few pieces of evidence we have for how veterinary medicine was practised during the hundreds of years between the disintegration of the Western Roman Empire and the first glimmerings of the Renaissance. This book was primarily written with human patients in mind, but it also makes numerous references to treating animals. I should point out that "leech" did not then, or for much of history, have the strongly pejorative connotations it does now. It simply meant "doctor" and was probably considered a complimentary, or at least honourable, title as it referenced what was considered a specialized tool that lay people did not feel they could easily master themselves. It's as if we called modern doctors "stethoscopes." As an aside, it is worth noting that Homer, back in 800 or 900 BCE, referred to doctors as "leeches," and that as recently as 1833, France imported 41 million leeches for medical purposes. So, this practice has a very long history. And moreover, leeches are back. I mean the little animals now, as they turn out to be useful in managing some wounds and for maintaining blood flow after certain plastic surgery procedures.

Icky, I suppose, but then ickiness is really just a matter of opinion and perspective, aspects of the human experience that have been fickle over the course of time.

The fictional Oswald in the foregoing story would have been called a horse-leech. The various kinds of leeches were self-trained individuals who developed a reputation for their skills in treating people or animals, or both. The use of the terms "horse-leech," "cow-leech," and, later, "dog-leech," implies that some became species specialists, at least by reputation. However, anyone could treat anything if they had access to the herbs, the tools, and, when indicated, the leeches. Unfortunately, there is very little in Bald's Leechbook which looks like it would have been helpful. The few exceptions are an eye salve, which has recently been shown to have good antibiotic properties, and a remarkably sensible method for amputating a limb. Otherwise, it is full of stuff like crab's eyes, rabbit bile, crushed ant eggs, and porpoise-skin whips. It sounds like a child's recitation of the contents of a witch's cauldron. Except the porpoise-skin whip. That's just weird.

In addition, there are remedies against night goblins (three stones from inside a young swallow's gizzard, in case you're curious) and a whole lot of treatments involving herbs mixed with holy water and then sprinkled on diseased animals. Taken internally some of these herbs might have had some effect, but the sprinkling is just a more theatrical version of prayer. And, of course, the Leechbook also spelled out that peculiar method for treating bloat in livestock caused by elves. I suspect that men like Oswald also suggested changes to diet and exercise that might help in some mild cases, and thereby scored a few more successes beyond those cases which would have improved on their own anyway. And when the animal remained "elf-shot" despite following the Leechbook's remedy to the letter? Oswald could always say that the owner did not call him quickly enough. Then, as now, there is an element of showmanship in inspiring a client's (or patient's) confidence that no amount of logic can replace.

As an aside, it's interesting to note that the elf-shot myth came about in part because farmers were constantly turning up small neolithic flint arrowheads when they ploughed. Archaeology not having been invented yet, and not having any other ready explanation, the story grew that wicked little elves were firing them at livestock. In some areas, the arrowheads were used in the curative ritual by, for example, rubbing them on the unfortunate animal's forehead, similar to how Oswald used his special knife.

In addition to herbs, holy water, and leeches, special talismans and amulets were also popular for their reputed medical power to ward off illness in man and beast. One of the most famous was the "Lee Penny" in Scotland. The story was that in the 14th century the crusader Sir Simon Lockhart captured a Saracen (Muslim) nobleman. The nobleman's mother, a wise Saracen healing woman, bought his freedom with a wondrous stone that she told Sir Lockhart was a cure for the bite of rabid dogs, and for many diseases of horses and cattle. Lockhart strikes the modern person as having been astonishingly credulous, but those were different times, and the stone apparently went on to gain fame in Scotland for its powers. It was eventually set in a silver fourpenny piece, hence the name. By the 17th century the Lee Penny was considered so powerful and valuable that it was granted a special exemption by the Church of Scotland, which otherwise banned charms and amulets as dangerous pagan delusions. At one point it was loaned to the city of Newcastle against a surety of six thousand pounds (a lot of money then) to tackle an outbreak of cattle plague. The city was so pleased by the effect that it offered to buy it from the Lockhart family, but they refused.

Another magical medical penny was the Lockerby Penny. According to William Henderson in his 1878 *Notes on the Folk-Lore of the Northern Counties of England and the Borders*, the Lockerby Penny was used to imbue otherwise ordinary water with the power to cure mad cattle.

> The Lockerby Penny is put in a cleft stick, and a well is stirred round with it, after which the water is bottled off and given to any animal so affected. A few years ago, in a Northumbrian farm, a dog bit an ass, and the ass bit a cow; the penny was sent for, and a deposit of 50 pounds actually left till it was restored. The dog was shot, the cuddy died,

> but the cow was saved through the miraculous virtue of the charm. On the death of the man who thus borrowed the penny, several bottles of water were found among his effects, stored in a cupboard, and labelled "Lockerby Water."

The Lee and Lockerby Pennies are only two examples. Magical coins, objects, and especially stones of all sorts have been pressed into veterinary service over the ages, including up to the 19th century, but their heyday was the Middle Ages. The literature is full of specific named stones, such as the Clach Ruaidhe, again in Scotland, which was a small red basalt rock that was reputed to cure mastitis (inflammation of a cow's udder) when rubbed on the affected area.

Words were also considered powerful. Numerous special incantations used by medieval cow-leeches have been preserved. The following is one that was used to treat "red water," or blood in the urine. The cow-leech collects the urine in his hands, throws it into the water, washes his hands, and then cups his hands to form a trumpet shape. Through this he chants:

> In the name of the Father of Love,
> In the name of the Son of Sorrow,
> In the name of the Sacred Spirit. Amen.
>
> Great wave, red wave.
> Strength of sea, strength of ocean,
> The nine wells of Mac-Lir,
> Help on thee to pour
> Put stop to thy blood
> Put flow to thy urine.
> (Name of animal).*

* Featured in T.D. Davidson's "A Survey of Some British Veterinary Folklore," *Bulletin of the History of Medicine*, 24, no. 3 (May–June 1960): 200–201, https://archive.org/stream/sim_bulletin-of-the-history-of-medicine_may-june-1960_34_3/sim_bulletin-of-the-history-of-medicine_may-june-1960_34_3_djvu.txt.

This is obviously a Christian gloss on a pagan incantation, but otherwise it is a remarkably straightforward plea for normal urinary function without a lot of the gobbledygook one imagines such chants to contain. I also find it interesting that it specifies that the animal be named. That it is a named individual with a specific identity tells us something about the relationship between peasant and livestock.

When "murrain" struck flocks of sheep in medieval England, it was common practice to gather the sheep to listen to a priest read the appropriate curative psalms to them. The word "murrain" was used as a generic term for infectious diseases among sheep and cattle, so what it referred to varied in severity and prognosis. Consequently, reading psalms may sometimes have appeared effective. And when it wasn't, it would have been easy enough to explain that the sheep had been presented too late, or the shepherd was sinful, or the day was inauspicious, or any one of an array of similar excuses. Apparent success only has one explanation, but failure, many. Regardless, it's hard not to take pleasure in the mental image of a priest reading from the Bible to flocks of sheep.

Meanwhile in Ireland, cow-leeches were allegedly practised in the art of shoving two live frogs down the throats of cattle afflicted with blood in the urine. This was to be preceded by six quarts of water and followed by three. The water probably helped a little bit, but otherwise, chanting seems more sensible than live frogs.

Back in Scotland again, cow-leeches advised farmers to protect their milking cows against disease by collecting bog violet, but only three tufts, and only from the tops of cliffs or mountains where no animals had trod, and only on a Sunday, and only by plucking, never cutting, and only while chanting a five-line incantation (I'll spare you) and then wrapping it in a cloth bag and tying it to the cow. There are many examples of such highly specific healing rituals, some far more elaborate than that. It is impossible to avoid the suspicion that the complexity was a safeguard against critique when the method failed. It would be very easy for the cow-leech to claim that the farmer had failed to meet some particular, highly specific, condition, and that the advice was otherwise sound.

In Shetland the custom was to drag a cat by its tail across the back of a cow while the cow was calving to safeguard the life of the cow and the calf. Safeguarding the life of the poor individual tasked with performing the cat-dragging appears not to have been a concern.

I could fill this entire book with the weird and wondrous ways of the medieval cow-, horse-, and dog-leech, but we have to move on.

The bottom line in all the foregoing was that the European Middle Ages, as in the ancient world that preceded it, saw very little progress in medicine, whether human or veterinary, though not for lack of trying. Animals were extremely valuable as individuals, far more so than they are in the West today. A single cow could represent most of a peasant's wealth. In the 14th century a good cow was worth about ten shillings at a time when a servant or a herdsman earned no more than ten shillings a year. Basically, the equivalent of the average cost of a new car today. Every effort would have been made to keep these animals in good health and treat their afflictions, but the knowledge was simply lacking. One of the most important factors in the pause of medical advances was the general belief that the world was not subject to change. God had ordered the world thus, and thus it would remain. "Research" in any form could be seen as a challenge to God's will. The church tolerated attempts by lay people to use folk remedies, many of which had pagan origins, but the overarching understanding was that disease was simply divine punishment for sin, more a matter of deserved fate than reversible poor luck. In a world where famines, plagues, and wars were frequent and where only about two of three children survived to see their tenth birthday and, as a result, the average life expectancy was 33, it was easy to believe that man was mostly helpless against his destiny. But, still, people tried. They called the leeches and the wise old women. They brewed up potions and poultices. They pounded herbs and chanted incantations. They prayed and made promises to God. There's no reason to believe that they didn't love their animals, despite the church's teaching that animals had no souls. But on the whole, if you had to be a cow, you would have been better off being one in Varro's Rome or, much better yet, Shalihotra's India, than in Oswald's England.

But it was not all darkness. Far to the south, in Salerno, Italy, glimmers of light began to appear in the 11th century when a medical school was founded that managed to evade the church's prohibition on dissection. It was a multicultural, multilingual, secular school where students could dissect pigs, which were known, then as now, to be surprisingly serviceable facsimiles of humans on the inside. A major figure at the school was the 11th-century physician, Constantinus Africanus, who was the first person to translate Arab medical texts into Latin. He recognized that Arab science was at that time well advanced over European. The great centres of learning in the Middle Ages were Baghdad, Damascus, and Cairo, decidedly not London, Paris, or Rome. This was in part because Islamic culture at the time was forward-looking and interested in progress as a concept, perhaps because it was a relatively new religion. It was also because they were eager to learn from the ancient Greeks and Romans, who had been rejected as pagan by Medieval Christian authorities. But more on Arab medical, and specifically veterinary, science in a moment. Let's wrap up the mostly sad state of affairs in Medieval Europe first.

Another point of light, albeit fainter, was the work of the famed mystic St. Hildegard of Bingen (1098–1179 CE). If you're old enough, you might recall that, seemingly out of the blue, her music suddenly became popular in the 1990s with dozens of CDs released of haunting chants and ethereal voices. The Enya crowd was delighted. More germane to our story, though, she also wrote two books on medicine which also covered animal diseases. One curious fact about these books is that she wrote them in a secret language using coded letters. It seems that medieval mystics had plenty of time on their hands. The books contain a lot of dubious incantations, but also many herbal remedies that remain in use today, such as stinging nettle. Hildegard recommended its external use in horses for injuries and inflammation, and although that wouldn't be our go-to today, in a pinch it wouldn't be wrong to try. Most of her recommendations were species-specific, such as marigold for cattle, oak leaves for goats, and crushed snail shells for swine. The latter, unsurprisingly, failed to gain any traction over time, but marigold and oak leaves do have medicinal properties. It is clear that Hildegard sought out existing folk wisdom regarding healing and paid attention to observable effects rather than giving advice based on blind faith alone. This was a step forward. Unfortunately, she also weighed in on the great rabies debate, adding yet another bizarre treatment to the list: lark's heads.

From Hildegard we jump ahead a full century to Emperor Frederick II (1194–1250 CE) of the Holy Roman Empire — a collection of mostly German and Italian lands that claimed to be the successor state to ancient Rome, but of which Voltaire quipped, "It is neither Holy, nor Roman, nor an Empire." That may have been mostly true, but during Frederick's time, it was important enough that he was one of the most powerful men in Europe. Fortunately, he was also one of the most curious and learned men in Europe and was commonly called "stupor mundi," or "wonder of the world," for his many skills and interests. He was also religiously unorthodox, possibly even a closet atheist, and was eventually excommunicated. Frederick took great pains to explain that he wanted to understand the world through observation and experimentation, thereby seeing things exactly as they are (ea que sunt, sicut sunt).

For the purposes of our story, however, it is his passion for animals, and especially falcons, that is of particular interest. His *De arte venandi cum avibus* (*The Art of Hunting with Birds*) is a medieval masterpiece, full of detail and betraying a keen sense of connection with the birds. It is impossible to know to what extent this interest was driven by a practical desire not to lose valuable animals and to what extent by affection for favourite individuals, but as I've stated before for others, it is reasonable to assume a bit of both. He consulted with Arab falconers, and I can clearly picture Frederick in my mind's eye: He sits in a sumptuously decorated castle tower, the Italian sun streaming in through the windows. Amidst this beauty and luxury, he looks sadly at his ailing gyrfalcon drooping on his perch. He asks the falconer if there's anything at all that could be done. In that moment nothing else matters to him. He's had some of his happiest moments with this bird, but now it's likely to die and, according to the church, become dust, nothing more. Frederick can't accept this. Sentimental speculation? Absolutely, but that doesn't make it wrong.

Around 1240 Frederick appointed Giordano Ruffo as Chief Imperial Horse Marshall and Veterinarian and directed him to compile as much information as possible about the care of horses. The result is one of the earliest books we can truly consider to be a veterinary textbook, the *De medicina equorum*. It is somewhat tainted by astrology but otherwise remarkably rational, or at least rational for the times. It lists 57 horse diseases and 150 medications and is especially noteworthy in recommending a gentle and kind approach to the patient to gain its trust, applying only the necessary amount of force for the purposes of restraint. It is also the first book to describe the ligation of veins to stop bleeding. As this is a fairly obvious solution, I suspect it had been in practice for a long time prior, but here it was in print finally. *De medicina equorum* was considered authoritative as late as 1818, when a new edition was published and sold well.

Ruffo also drew heavily from Arab sources, so it's time we give them their due, before turning the page to the Renaissance.

Within a hundred years of Mohammed's death in 632 CE, Islam had spread from its origins in the Arabian Peninsula to a vast territory from Spain in the west to Afghanistan in the east, and this spread was on horseback. Horses were venerated as a kind of cult animal by the Arabs. They were accorded a special status in the culture and society. There was no comparable human-animal relationship in the Christian West. And as mentioned earlier, Islamic leadership encouraged borrowing from the ancient Greeks and Romans, so, as their warriors overran territories once ruled by Rome or where Greek culture had been active, they busied themselves with translating the ancient texts into Arabic and then seeing how they could build on those foundations. Hippocrates was revered, and a modified version of his oath was used for physicians and veterinarians throughout the Islamic world. Aristotle was also a great favourite, although Arab writers such as Ibn Sina quickly improved on him.

Ibn Jakoub, official veterinarian to the Caliph, wrote a book in 695 CE that drew not only from Greek veterinary sources, but Persian and Indian as well, since the Islamic Empire had pushed that far east. A few hundred years later, Moshe ben Maimon, a Jewish* rabbi, philosopher, physician, and veterinarian working in Cairo, made the breakthrough observation that rabies was transmitted through a dog's saliva. He called it "the most dangerous of all poisons." Hurray! No more tongue worms or lark's heads! Unfortunately, it took a long time for this information to filter up to Europe. Ben Maimon also studied tuberculosis in cattle in slaughterhouses and should be regarded as one of the pioneers of veterinary public health.

Although in terms of timeline he overlaps with the unfolding Renaissance to the north in Europe, Abu Bakr ibn Badr al-Din al-Baytar (1309–1340 CE) is worth a mention. Incidentally, "baytar" is the Arabic word for "veterinarian." He was the horse master to the Sultan in Cairo and wrote extensively on veterinary matters, again borrowing from, and improving on, the Greeks and on Vegetius (remember him?). In particular, he advocated methods of treatment for fractures in horse legs that would not be wildly out of place today. He also innovated the use of a copper tube to drain fluid from a horse's belly.

To tell the global story of the history of veterinary medicine, one might think that I should spend more time in the Islamic world, and go back to India and China as well, but for two reasons I won't. The first is that after a promising start, innovation in medicine, veterinary or otherwise, slowed to a trickle in those parts of the world. By the end of the Middle Ages, the action did shift to Europe. And second, most of the rest of this story takes place in Europe, and later North America, because the line

* It is worth pointing out that the Islamic world was comparatively tolerant of other religions during the Middle Ages, in contrast to the situation for religions other than Christianity in Europe.

that your veterinarian stands in, whether they are in Nairobi, Toronto, Paris, Hong Kong, Mumbai, Dallas, Buenos Aires, or Cairo, can be traced directly back to that first modern school in Lyon.

We're on our way, but still many hundreds of years from Lyon. Before we get there, we need to make a few stops, beginning with a hunting lodge in the far southwest corner of France in the 14th century.

CHAPTER SEVEN

Fébus

Gaston sat on the bench in the front hall and pulled his boot off with a great deal of effort, groaning as he did so. He looked balefully at the other boot and took a breath before attempting it. They were of the finest Italian leather. They attracted nothing but admiring comment. But they were new, and they were tight. The long day's ride with the hounds had caused his feet to swell. And they hadn't even caught the stag. It had not been a good day. To compound matters, Belle, his best bitch, had been off her game. She kept stopping to scratch herself. Her ears and neck were bloody by the end of the day. And the November rains were cold and miserable. It really had not been a good day.

"Sir, allow me to help you." Tomas bent down and reached for the boot.

Tomas still had a faint Spanish accent. He was young, maybe 14, and had come to Foix as a small boy from over the Pyrenees in Navarre with Gaston's wife's retinue. Of all Gaston's attendants and servants, Tomas was the most attuned to his needs, but Gaston waved him off. "I can manage," he said. He grunted as he laboured with the boot. "Get Belle for me. I need to look at her."

Eventually the boot came off. While Tomas fetched Belle, the other servants lit lamps and stoked the fire in the adjoining great chamber. One of them handed Gaston a glass of wine as he sat in his favourite chair facing the hearth. He was joined by his cousin Robert, who had stayed behind because of an injury sustained when he fell off his horse on the previous day's hunt.

"How was it today, Fébus? Better than yesterday?" Robert asked. Fébus was Gaston's nickname — a reference to Fébus (Phoebus) Apollo, the sun God, because of Gaston's long golden hair. No matter that it was matted and dirty at the moment.

"No, not better. Worse." Gaston laughed and took a long deep swallow from his glass. "That stag is the devil himself. I saw him once through the rain, on the ridge by Saint-Faust, but then he vanished. The hounds could not find his scent. Too much rain. Too much devilry."

"Not even Belle?"

"No, not even her. And there is something wrong with her." Gaston looked over his shoulder at the door. "Speaking of which, where is Tomas? He was to bring her to me."

As if on cue, he heard a scrabble of nails on stone in the front hall, and Tomas entered the great chamber with two wet dogs on leather leads. He let the leads go. "Go see your master."

"Ah, you've brought me Philippe too?" Gaston clapped his hands to encourage the dogs to come to him. Belle and Philippe ran across the room, tails wagging. They were long-legged, long-nosed hounds, about knee-high. Belle had black, brown, and white markings, while Philippe was mostly brown with a white blaze on this chest.

"Yes, sir. I hope that is all right. He is also scratching at his ears, although not as badly as Belle."

"Do you not want the farrier or the village dog-leach for this, cousin?" Robert asked.

Gaston snorted and shot Robert a dark look. His cousin loved to provoke him. He knew full well what Gaston thought of farriers and leeches. He turned back to the boy and the dogs.

"Well observed, Tomas. I did not see that myself. But you are correct." Gaston bent down to look at Philippe's ears. "Bring me some more light."

Tomas brought a candelabra over from a side table. Robert watched with a bemused expression while swallowing the last of his wine. He snapped his fingers for a servant to refill his glass.

"Philippe is suffering from the rheum," Gaston said and straightened up.

"The rheum, sir?"

"Yes, a cold in the ears. The weather brings it on. He will scratch himself until he is deaf."

"That's horrible sir. Is there no remedy?"

Gaston laughed. "Yes! Happily, there is an excellent remedy. Wash his ears out with lukewarm wine on a cloth. Do this four times a day. After each washing, put in three drops of camomile in oil. Make sure that he does not scratch! You will have to watch him constantly. In a few days, he will be well. Can you do this?"

"Yes, sir. And what of Belle?"

"Belle is a greater worry. Let Philippe lay by the fire and bring Belle into the light."

"She does not look well, Fébus," Robert said when Tomas held the candles over her. "Look how she bleeds."

"She has done that to herself with her scratching. It is a mange, I am sure. But which one, I do not know. I must look at her carefully."

"A mange? Not farcy?" Robert asked, leaning forward.

"No, with farcy the glands swell like walnuts, like glanders in horses. Her glands are not swollen."

"I defer to your greater knowledge, cousin," Robert said, and sat back in his chair, making a sweeping cap-doffing gesture with his left hand.

Everyone was quiet for a few minutes while Gaston peered at Belle's skin in the flickering candle- and firelight. The only sounds were the crackle of the fire and Philippe licking his paws.

"This is the flying mange," Gaston said.

"Flying?" Tomas asked. "From flies or other small, winged things?"

"We do not know where it comes from, but it goes from dog to dog as if flying. But for this I have a remedy as well."

"Are you going to spend more wine on your hounds, cousin?" Robert said and smiled. "Wine in the ears, wine on the face. I only put wine where, according to God, it belongs." He held his glass to his lips and then, realizing it was empty, raised his eyebrows and snapped his fingers.

"No. Something much stronger is needed. Listen well, Tomas, so that you may learn. But I will come and apply the remedy myself after my supper. It is not for boys to trifle with."

"Yes, sir." Tomas said, face serious. Belle lay down at Gaston's feet.

"First, you take quicksilver and make an ointment of it by putting it in a dish with the spittle of three or four fasting men —"

"Ha!" Robert interrupted. "Where shall you find such men in this castle? Not I, I'm afraid."

"The servants' supper will be later tonight," Gaston said with a slight shrug. "Stir the quicksilver and spittle together against the bottom of the dish with a pot-stick until well mingled. Then take as much verdigris as of the quicksilver and mix it in, always stirring with a pot-stick. Next, take a great piece of old swine's fat without salt, and take away the skin above. Put this in the dish with the mixture and stamp it all together for a long while. With this I will anoint Belle, but only where she has mange and nowhere else. After, you must watch closely and be very sure that no other hound tries to lick it, for it cannot be taken internally. She will be unable to lick it herself as the mange is only about her head and neck."

"I understand, sir, thank you. If I may ask, though, what is verdigris?"

"It is copper that has become blue-green in the air."

"Here's a toast to you, cousin," Robert said, raising his glass. "You are more learned in the ailments of hounds than any man I know. Even more than that wretched dog-leech." He winked at Tomas.

"Thank you, Robert. I will take that as a compliment, although you have set the comparison very mean." He laughed, picked up his own glass, and raised it to his cousin. "After a man's horse, his hunting dog is his most treasured companion."

"More treasured than his mistress?" Robert asked, and grinned.

"Far more. Mistresses are much easier to replace. And such it behooves him to do everything in his power to keep his dog well."

Robert nodded. "Then, cousin Fébus, you must write a book so that your knowledge is not lost when you fall from your horse and strike your head."

"Ah, but that is *your* special talent, Robert."

The two men burst into laughter, startling the dogs. Tomas smiled, but stayed quiet, trying to remember whether it was four washes with wine and three drops of camomile or three washes with wine and four drops of camomile.

Gaston Fébus, the 11th Count of Foix, did go on to write the *Livre de chasse* (*Book of the Hunt*) between 1387 and 1389. Its 36 chapters have titles such as "Of the Wild Boar, and of His Nature" and "How a Man Should Know a Great Hart by the Fumes" (fumes meaning feces), but there are nine chapters on the care of dogs, including one specifically on the medical aspects, "Of Sicknesses of Hounds and of Their Corruptions." Up until this point in the history of the veterinary arts, scarce attention was paid to any species beyond horses and cattle (and elephants in India). Dogs were typically only mentioned in the context of rabies. Fébus, however, clearly loved his dogs. *Livre de chasse* is filled with affectionate language for them and descriptions of the joy they had together in the hunt. His last words, before dying peacefully in bed after a long day's hunt, were apparently of which hounds had that day run the best and had the best noses.

Livre de chasse proved to be very popular and was translated into English 20 years later as *The Master of Game* by the Edward of Norwich, the second Duke of York. Edward would go on to fall at the Battle of Agincourt, where he was unfortunately "smouldered to death by much heat and pressing."* *The Master of Game* circulated for centuries in manuscript form but was not printed until 1909 when a popular edition was produced with a lengthy foreword by the American president, and noted hunter, Theodore Roosevelt. Roosevelt had kind words for Gaston and Edward:

* Other claims to fame include appearing as two different characters in Shakespeare's works: Edward of Langley, the Duke of Aumerle, in *Richard II*, and, later, the Duke of York in *Henry V*, where his death at Agincourt is made to look far more chivalrous and stirring.

> Both the Count of Foix and the Duke of York show an astonishing familiarity with the habits, nature, and chase of their quarry. Both men, like others of their kind among their contemporaries, made of the chase not only an absorbing sport but almost the sole occupation of their leisure hours. They passed their days in the forest and were masters of woodcraft.

But Roosevelt does not comment on Fébus's medical advice. Nor has anyone else, but in fact some of the remedies he recommended are not completely out of line. It's easy to look down on our forebears from our high 21st-century perch, but they weren't making capricious or random medical decisions. A great number of empirical observations over many years went into determining which treatments worked and which didn't. For men such as Gaston Fébus, these dogs were very valuable, having both practical and emotional importance. It was well worth their while to put effort into giving the best medical care contemporary knowledge could provide.

For example, while we have much more effective treatments now, wine and camomile are reasonable options given what was available in 14th-century France. In *Livre de chasse*, Fébus describes the "rheum" in the ears like this: "They claw themselves so much with the hinder feet that they make much foul things come out thereof."

This sounds an awful lot like the common yeast and bacterial overgrowths we see in dogs' ears which have become inflamed due to allergies, trauma, or swimming. If you've ever smelled such an ear, you'll agree with the "foul things" description. Wine has antiseptic properties and will lower the pH in the ear canal. Camomile is also mildly antiseptic.

The same goes for the "flying mange" treatment. Mercury and copper are far too toxic to contemplate using today, and Fébus clearly understood the hazards, but they would work to kill mange mites. The spittle of fasted men is obviously less useful. The origin of these recipes would have been unknown to their users. They were handed down over generations. It would not have been obvious what the mechanism of action was, or which elements were most important, so a policy of not messing with what had always worked was prudent. Fasting spittle didn't hurt, and maybe it helped, so why not?

The Master of Game is a delight to read today. I cannot vouch for the original French, but there is a charm to the English of Edward's time. Also, unlike the great majority of previous texts on veterinary matters, Fébus comes across as a true dog lover, a sensible empiricist, and ultimately a very practical man. Take this passage, for example, where he discusses care of their paws and nails:

> Also many hounds be lost by the feet, and if some time they be heated take vinegar and soot that is within the chimney, and wash his feet therewith until the time that they be whole, and if the soles of the feet be bruised because, peradventure, they have run in hard country or among stones, take water, and small salt therein, and therewith wash their feet, the same day that they have hunted, and if they have hunted in evil country among thorns and briars that they be hurt in their legs or in their feet, wash their legs in sheep's tallow well boiled in wine when it is cold, and rub them well upward against the hair. The best that men may do to hounds that they lose not their claws is that they sojourn* not too long, for in long sojourning they lose their claws, and their feet, and therefore they should be led three times in the week a-hunting, and at the least twice. If they have sojourned too much, cut ye a little off the end of their claws with pincers ere they go hunting, so that they may not break their claws in running. Also, when they be at sojourn, men should lead them out every day a mile or two upon gravel or upon a right hard path by a river side, so that their feet may be hard.

He was also, understandably, greatly concerned with their reproductive health, as breeding excellent hunting hounds was of particular interest. This

* In this context, "sojourn" means to stay at rest without exercise.

results in one of the more interesting sections. For this you will need to know that a "yerde" is the penis, and "to scombre" is to defecate. I suspect that I do not need to translate "ballock purse" for you.

> Sometimes an evil befalls in the ballock purse, sometimes from too long hunting or from long journeys, or from rupture, or sometimes when bitches be jolly, and they may not come to them at their ease as they would, and that the humours runneth into the ballocks, and sometimes when they be smitten upon in hunting or in other places. To this sickness and to all others in that manner, the best help is for to make a purse of cloth three or four times double, and take linseed and put it within, and put it in a pot, and let it mingle with wine, and let them well boil together, and mix it always with a stick, and when it is well boiled put it within the purse that I spoke of, as hot as the hound may suffer it, and put his ballocks in that purse, and bind it with a band betwixt the thighs above the back, make well fast the ballocks upwards, and leave a hole in the cloth for to put out the tail and his anus, and another hole before for the yerde so that he may scombre and piss and renew that thing once or twice until the time that he be whole. Also, it is a well good thing for a man or for a horse that hath this sickness.

Good to know that this works for humans as well. That alarming suggestion aside, once again, Fébus gives reasonable advice, given the times and the available medicines. The "evil" befalling the "ballocks purse" is likely infection, resulting in a painful, swollen scrotum. Wine, again, has antiseptic properties, and heat can be used to draw out infection. It is good to see that he is mindful to make it only as hot as the dog will tolerate.

However, like every proto-veterinarian before him, and everyone after for another five long centuries, Fébus was completely at sea when it came to understanding and treating rabies. We will look at the rabies story in greater detail later, but it is interesting to note the state of the art in 1389:

> The remedies for men or for beasts that be bitten by mad hounds must need be done a short time after the biting . . . Some goeth to the sea, and that is but a little help, and maketh nine waves of the sea pass over him that is so bitten. Some take an old cock and pull all the feathers from above his vent and hangeth him by the legs and by the wings, and setteth the cock's vent upon the hole of the biting, and stroketh along the cock by the neck and by the shoulders because that the cock's vent should suck all the venom of the biting. And so men do long upon each of the wounds, and if the wounds be too little they must be made wider with a barber's lancet. And many men say, but thereof I make no affirmation, that if the hound were mad, that the cock shall swell and die, and he that was bitten by the hound shall be healed. If the cock does not die it is a token that the hound is not mad. There is another help, for men may make sauce of salt, vinegar and strong garlic pulled and stamped, and nettles together and as hot as it may be suffered to lay upon the bite. And this is a good medicine and a true, for it hath been proved, and every day should it be laid upon the biting twice, as hot as it can be suffered, until the time when it be whole, or else by nine days. And yet there is another medicine better than all the other. Take leeks and strong garlic and chives and rue and nettles and hack them small with a knife, and then mingle them with olive oil and vinegar, and boil them together, and then take all the herbs, also as hot as they may be suffered, and lay them on the wound every day twice, till the wound be healed, or at least for nine days. But at the beginning that the wound be closed or garsed [cupped] for to draw out the venom out of the wound because that it goeth not to the heart. And if

> a hound is bit by another mad hound it is a good thing for to hollow it all about the biting with a hot iron.

It is good to see Fébus the empiricist at work here again, commenting on what he has seen work and what is just hearsay. Of all the foregoing, only the very last thing he says, cauterizing the wound, has some merit. So-called St. Hubert's* keys were used throughout Europe to try to prevent rabies in humans up until the early 20th century. With an appearance more like large nails than keys, priests kept them and would be called upon to heat them red hot and press them into the bite wound when rabies was suspected. Performed immediately after the bite, and thoroughly enough, it does stand a chance of killing all the rabies virus particles. In Fébus's case with dogs, the suffering of the poor animal being "hollowed all about the biting with a hot iron" can scarcely by contemplated, but then the suffering of dogs stricken with rabies is even more extreme. Euthanasia would be kindest, but what would you do if it was your beloved and prized hunting hound?

Although veterinary medicine may not have made any great advances during the Middle Ages, there is a wealth of fascinating information about its practice by a wide range of people. It would be a shame to leap directly into the Early Enlightenment and not make another stop. So let us move ahead 150 years to 1535, and across the Channel to the verdant hills of Derbyshire, where two shepherds are in debate with each other.

* One of several saints associated with protection against rabies.

CHAPTER EIGHT

FitzHerbert

"It is surely the pox," Tom said. The tall, young redhead leaned on his staff and looked at the small flock of 20 or so black-faced sheep just beyond the low stone wall.

"It is surely not," replied Edmund, his shorter, older companion. It was raining lightly, but neither man seemed to notice or care. Nor did the sheep.

"If you are wrong, Edmund, you will pay for it. His lordship will not be pleased. Many could die. And you will not blame me. I will tell him that I thought it was the pox and that you prevented me from trying to save them."

Edmund sighed and kicked at the grass. "Fine. It is not the season for the pox. Nor have his lordship's flocks ever been affected. Nor has it been seen in these parts for many years. But if it will stop you from prattling on about it, bring me the ewe you are so exercised about. And be quick about it, Tom. We still have to bring the others up from the river pasture before dark."

Tom trotted over to the wall, found the small protruding slabs of slate that served as footsteps, and clambered over. The sheep were used to him, so they didn't immediately look up from their grazing. However, there were nervous glances and shuffled hooves as he began to move among them.

He located the ewe he was looking for and picked her up. Normally he would have herded her with his staff towards the gate 50 yards to the south, but Edmund was impatient today. It was quicker to carry the sheep and heave her up over the wall for Edmund to have a look. The sheep submitted quietly to this indignity.

Panting slightly from having run and carried the ewe, Tom said, "Look at her withers. Are those not the marks of the pox?"

Edmund bent over, squinted, and parted the wool. There were several raised red bumps, each about the size of a farthing. He bit his lip. Admitting he was wrong was going to be hard to bear. Especially admitting it to Tom. But being dismissed, or worse, by Lord FitzHerbert would be harder to bear.

He straightened up and scratched his chin. He looked at the sheep for a long moment. "By the bones of holy St. George and St. Drogo, that does look like the pox. I saw it once before, many years ago. It was terrible." He was not going to apologize, but he couldn't staunch his curiosity about something. "How do you know what the pox looks like in sheep?"

Tom didn't try to hide his pleasure at not only being right, but also at being recognized as right. He beamed and then cleared his throat and wiped the rain from his face. It was raining harder now. "I learned it from Master FitzHerbert's book."

"You can read?" Edmund asked, not making any attempt to hide his incredulity. "I have never seen you read so much as a psalm."

"No, I cannot, but his Lordship has read parts to me."

Edmund grunted. His first thought was to wonder, not a little bitterly, why Lord FitzHerbert would favour Tom in this way. But he immediately realized that it simply meant that his lordship trusted that he, Edmund, already had all the necessary knowledge. Tom was just a whelp and needed as much learning and help as possible. Sheep were not in his blood. Edmund, on the other hand, had learned to be a shepherd at his father's knee, who had learned from his father before him, and so on.

"I was in the big house when I was a boy," Tom went on when Edmund didn't say anything. "His lordship wanted to hear how his words sounded when read aloud, and he thought I would benefit from some of the learning."

"I have seen the book. It is a fine one," Edmund said gravely. He couldn't read either. He remembered his father's remedy for the pox, although he couldn't recall the exact proportions of the herbs and roots involved. Perhaps Master FitzHerbert's book contained these details. Give him the rot, or foot worms, or wood-evil, and Edmund was confident in his remedies, but he hadn't seen the pox in so long, and it was a very serious sickness. Also, his lordship was now regarded as an expert in husbandry throughout the land. It would be prudent to consult with him first and have him read from his book.

Edmund sent Tom down to the river to bring in the rest of the flock while he headed back along the muddy lane to Norbury Manor, seat of the FitzHerbert family since shortly after the Norman Conquest. He was met at the door by stoop-backed old Percival, Sir Anthony's steward, who ushered him into the great hall after sharply telling him to take off his wet coat and wipe his boots.

"His lordship will be with you momentarily," Percival said before bustling off, leaving Edmund to sit on a wooden bench, gazing up at the portraits of dead FitzHerberts, including Sir Ralph, Sir Anthony's father, whom Edmund remembered well. It was cold as they had not set a fire in the enormous hearth at the end of the hall. Edmund assumed it was thought to be too early in the season, and perhaps Lady Mathilda, who was known to enjoy her comforts, was away.

A side door opened, and a large man entered. Although in his mid-sixties now, Sir Anthony had the erect posture and bearing of a much younger man. He was well dressed in a stylish bright green velvet tunic over deep red breeches. Edmund knew him well enough not be afraid of him, but others found his size and stern look imposing. Edmund stood up and bowed.

"Edmund," FitzHerbert boomed. "How are my sheep?" Sir Anthony FitzHerbert, Lord of Norbury Manor, King's Sergeant-at-Law, respected across the realm as a senior judge, and author of many renowned books, was not one for small talk. Certainly not with the servants and farm help.

"Most of them, by God's grace, are well, my Lord, but there is one who troubles me."

"Only one? It must then be of especial concern for you to come to me with this news. For a single sheep's fate, although no doubt of great import to that sheep, is of little interest to me." FitzHerbert smiled as he said this, making it clear that he was confident that Edmund would not be so foolish as to waste his time.

"Yes, of especial concern sir. It is a fine yearling ewe, and I believe it is suffering from the pox."

"The pox? Are you sure? It has not been seen in Derbyshire since my father's time."

"I am as sure as I can be, sir. The marks are raised and red and are the size of a farthing."

FitzHerbert sat heavily on a bench. "Sit. Tell me more."

Edmund described his examination of the ewe in detail, leaving out Tom's role.

"And it is the only one?"

Edmund paused a half second, realizing that he had not thought to examine the rest of the flock before coming up. But Tom would have, and he would have said something. Surely, he would have. "Yes, sir, I believe she is the only one."

"This is the flock up by Roston?"

"No, sir, thankfully the smaller one by the River Dove, and that flock is split in two. The poxy ewe is in with only 20-some of her fellows."

"Ah, there we are fortunate. The pox spreads quickly and kills like the plague."

"Which salve shall I prepare sir? My father always favoured —"

FitzHerbert cut him off. "No salves, no ointments, no balms. No shepherd's remedies, Edmund. None is of any use against the pox. And worse, they waste our time." Edmund was astonished, but he knew better than to argue. This was the author of the book, after all. And regardless, this was his lord and master.

"Take that boy Tom and inspect all of the flock carefully. Tom is clever and has a sharp eye. Check every part of their bodies, and if you find the slightest sign of the pox, separate that sheep from the others, wash him, and put him on fresh grass. Do this every day and keep separating any with signs. By this means we will halt the spread, and the new grass has some chance to cure the afflicted."

"Yes, sir. We will begin tonight." Edmund remained skeptical but the lord's method at least had the advantage of not making him have to scour the countryside for various herbs and roots, and then cook up a stinking mess of them. He remembered this well from his father's time.

"We will lose some, but we will not lose the whole flock. Take particular care with the lambs as they are most vulnerable."

Edmund nodded and waited to see if there were further instructions.

"Well, man, the flock will not sort itself into the poxy and the clean! Make haste. And pay heed to Tom. He will know which are afflicted."

FitzHerbert grinned at Edmund. It was as if he knew. But Edmund had no idea how he could.

Sir Anthony FitzHerbert (1470–1538) was yet another one of those polymaths we frequently encounter in the history of veterinary medicine. The book referred to above is his *Boke of Husbandry*, first published in 1523, but he is known better for his work in jurisprudence. He was a well-respected judge and published the first ever summary of English law, *La graunde abridgement*, in 1514. His subsequent legal text, *La novelle natura brevium*, is still cited today in common-law judgments. He followed the *Boke of Husbandry* with another perspective on rural life and economy, *The Boke of Surveying and Improvements*, but for our purposes it is *The Boke of Husbandry* that is of greatest interest. It contains a good summary of what we can assume was the state of the art for keeping, feeding, and breeding farm animals. Veterinary practices were considered integral to husbandry. A good cattleman or shepherd should have a grounding in the diagnosis and treatment of animal diseases. Cow-leeches, and presumably sheep-leeches (although I haven't run across that term), still roamed the land, but were primarily hired by peasants with a single cow or a couple of sheep. The gentry with their larger herds and flocks, and anyone with the means to purchase *The Boke of Husbandry* (and ability to read it), were better served by the DIY wisdom contained therein.

Much as in ancient times, throughout the Middle Ages veterinary medicine was a hodgepodge, with a few sound practices mixed into a breathtaking range of unsound ones. For example, FitzHerbert's approach to sheep pox, described in the story above, is perfectly sensible. In fact, things haven't changed much for that disease in the intervening five centuries. We still don't have an effective specific drug or remedy, and we still treat primarily through husbandry methods similar to what is described in the "*Boke*." But this didn't represent an advance on FitzHerbert's part. You might recall that Varro, back in Rome around 50 BCE, already recommended separating sick animals from the herd. The distribution of useful knowledge was spotty right up to the 18th century. It was impossible for an individual animal owner, whether a peasant or a lord, to get an objective sense regarding what was likely to work and what was not. It was all highly subjective and highly fragmented. There was no widely agreed upon unified body of knowledge. And there was no unified profession.

Human medicine was bedevilled by similar issues. In England, for example, although the College of Physicians of London was founded in 1518 — coincidentally, exactly when Sir Anthony FitzHerbert was writing — it wasn't until 1858's *Medical Act* that the force of law was applied to weed out unqualified practitioners. It took another 90 years, until 1948, for full exclusivity over the treatment of animals to be granted to licensed veterinarians in the UK (it was a little earlier in Canada and the USA, but not much).

But back to old FitzHerbert. Quite possibly the most astonishing passage in the *Boke of Husbandry* is in chapter 62, "The Turne, and Remedy Therefor." By "the turne" he's referring to a condition where a cow is showing central nervous system symptoms, including walking in tight circles. Read on for the detailed description and the amazing treatment suggestion:

> There be beastes that wyll turne about, whan they eate theyr meate, and wyll not fede, and is great jeoperdy for fallynge in pyttes, dyches, or waters. It is bycause that there is a

> bladder in the foreheed bytwene the brayne-panne and the braynes, the whiche must be taken out, or els he shal never mende, but dye at lengthe, and this is the remedy and the greatest cure that can be on a beaste. Take that beast, and cast him downe, and bynde his foure fete together, and with thy thombe, thrust the beast in the foreheed, and where thou fyndest the softest place, there take a knyfe, and cut the skyn, three or foure inches on bothe sides bytwene the hornes, and as moche benethe towarde the nose, and fley it, and turne it up, and pyn it faste with a pyn, and with a knyfe cut the brayne-pan two inches brode, and thre inches longe, but se the knyfe go no deper than the thycknes of the bone for perysshynge of the brayne, and take away the bone, and than shalt thou se a bladder full of water two inches longe and more, take that out, and hurte not the brayne, and thanne let downe the skynne, and sowe it faste there as it was before, and bynde a clothe two or thre folde upon his foreheed, to kepe it from colde and wete 10 or 12 dayes. And thus have I seen many mended. But if the beaste be fatte, and any reasonable meate upon hym, it is best to kyll hym, for than there is but lyttell losse. And if the bladder be under the horne, it is past cure. A shepe wyll have the turne as well as a beast, but I have seen none mended.

Yes, gentle reader, medieval brain surgery on cattle. It staggers the imagination to picture the poor animal being restrained by the simple expedient of "cast him downe, and bynde his foure fete together" before proceeding to find the softest spot in the skull with your thumb and then cutting away, taking care not to go too deep, lest there be "perysshynge of the brayne." Wouldn't want that.

Modern experts believe that the "bladder" he referred to was likely something called a hydatid cyst. These are produced by the tapeworm *Echinococcus granulosis*. It remains a worldwide problem today, albeit rare; it appears to have been far more common in the Middle Ages. You'll be excited to hear that it occurs in humans too, usually among poor rural children in the Global South. The great majority of these involve the liver and lungs, not the brain, if that is any comfort to know. Surgery remains the treatment of choice, in people and in cattle, although the objective now is to remove the entire cyst, not just drain it. Thankfully, anaesthesia has come a very long way since the 1500s.

FitzHerbert is also refreshingly frank when he doesn't have an answer, as is noted above with sheep who "turne," where apparently brain surgery is not an option. In another example, for gout he says, "I knewe never manne that coulde helpe it, or fynde remedye therfore, but all-onely to put hym in good grasse, and fede hym." This is far better than the random torments inflicted by other healers in vain attempts to address diseases they didn't understand. Not to say that FitzHerbert was above torments. For "warrybrede," which according to Middle English dictionaries is either a wound or a parasitic worm, possibly a bot larva, one is to again "cast hym downe, and bynde his foure fete together," and then:

> Take a hot iron, and sear it, take a culture, or a payre of tonges, or such an other yren, and take it glowing hote: and if it be a longe warrybrede, sere it of harde by the body, and if it be in the beginninge, and be but flatte, than lay the hot yren vpon it, and sere it to the bare skyn, and it will be hole for ever.

Ouch. Mind you, this is not that far off branding, which of course is still enthusiastically practised today.

He was also greatly concerned with "rotten" sheep and writes out a detailed list of symptoms, from the appearance of the white of the eye to how easily the wool is pulled off, by which one can determine rottenness of a sheep. He concludes this passage with a recommendation to inspect the liver of a slaughtered sheep suspected of being rotten:

> Also whan thou haste kylde a [rotten] shepe, his belly wyll be full of water, if he be sore rotten, and also the fatte of the fleshe wyll be yelowe, if he be rotten. And also if thou

> cut the lyver, therin wyll be lyttell quikens lyke flokes, and also the lyver wyll be full of knottes and whyte blysters, yf he be rotten; and also sethe the lyver, if he be rotten it wyll breke in peces, and if he be sounde, it wyll holde together.

FitzHerbert is describing parasitic liver flukes and is once again pretty accurate. Veterinarians working in slaughterhouses still look for the same things.

However, before we praise the old fellow too much, it should be noted that chapters 71 through 78 list the properties of various animals, from horses and cattle through, curiously, to badgers and foxes, until chapter 79, which is titled "The Properties of a Woman." Then, without breaking stride, he goes blithely on to describe diseases in horses in chapter 80 and beyond. Maybe he thought he was being droll. Maybe we now know why Lady Mathilda was away.

There were a few more veterinary-related publications in the 1500s in Italian and French, but none moved the practice forward. In fact, Federico Grisone, writing in 1550, achieved a new low in bizarre cruelty when he advocated tying a hedgehog under a recalcitrant stallion's scrotum to stop the horse from running backwards. He also states that a difficult horse should be punished by strapping a cat to its underside. That this also punishes the cat seems to have escaped his notice.*

The next major English language work relating to veterinary medicine was *Markham's Maister-Peece*, written by Gervase Markham's in 1610. He means "masterpiece," in case that's not immediately obvious. Markham (ca. 1568–1637) was not a humble man. The full title, if you're ready for it, was:

> Markhams maister-peece, or, What doth a horse-man lacke

* All this and more, is in Grisone's *Gli ordini di cavalcare* (*The Rules of Horsemanship*). Unfortunately, he was very influential in his time.

> containing all possible knowledge whatsoeuer which doth belong to any smith, farrier or horse-leech, touching the curing of all maner of diseases or sorrances* in horses : drawne with great paine and most approved experience from the publique practise of all the forraine horse-marshals of Christendome and from the private practise of all the best farriers of this kingdome : being divided into two bookes, the first containing all cures physicall, the second whatsoever belongeth to chirurgerie [surgery], with an addition of 130 most principall chapters and 340 most excellent medicines, receits and secrets worthy every mans knowledge, never written of nor mentioned in any author before whatsoever : together with the true nature, use, and qualitie of everie simple spoken of through the whole worke : reade me, practise me, and admire me.

Whew.

"Admire me." Indeed. One cannot help but admire his brashness. In wading through these 130 "principall chapters," I was hard-pressed to find anything of value. Casual misogyny aside, FitzHerbert was easily the better source. Again, a demonstration that the distribution of helpful medical knowledge was random and patchy, not moving forward with the centuries, but lurching about, much as a drunk does who occasionally utters something wise. Markham primarily drew from traditional farrier practices. We'll address the conflict between lay farriery and professional veterinary doctors later, but suffice it to say that farriers, while well versed for the times in the care of horses' feet, relied on word-of-mouth folk wisdom, handed down over the generations, when they strayed from the hoof.

Markham's "most excellent medicines, receits [recipes] and secrets" included such oddities as making a horse with "frenzy" sneeze by fuming his nostrils with the smoke of burnt garlic, or taking a "live pidgeon, and

* Sorenesses.

cleaving her in the midst, lay it hote unto the wound." Apparently, this will "draw out the venome." Once the unfortunate pigeon's work is done, one is to "heale the soare with turpentine, and hogges grease well molten together." He was alarmingly keen on this idea of cleaving smaller animals in two as part of a treatment. He recommended — *sensitive readers be forewarned* — cutting the head and tail off a cat and then splitting her in two and clamping her, "while still hot," around a horse's strained sinew. The mind boggles.

Later in the 17th century we find a farrier, Andrew Snape (1644–1708), who made a positive contribution to veterinary medicine. His 1683 *The Anatomy of an Horse* was wonderfully illustrated and remarkably accurate for the time. The full title is much more modest than Markham's, signalling more sober content:

> *The Anatomy of an Horse containing an exact and full description of the frame, situation and connexion of all his parts, (with their actions and uses) exprest in forty-nine copper-plates): to which is added an appendix, containing two discourses, the one, of the generation of animals, and the other, of the motion of the chyle, and the circulation of the bloud.*

Snape was Serjeant Farrier to King Charles II, meaning that he was the chief farrier in the land, only the second person to hold that position. Unlike most of his brethren, who plied their farrier's trade using only tradition as a guide, Snape appears to have been able to read Latin and made use of Renaissance Italian sources for his book. Among his 101 chapters, he included details on the function of the pancreas to secrete into the digestive system, and a description of roughly how the brain works:

> It is generally agreed that the proper action of the Brain (taken in a large sense) is the elaborating of Animal Spirits; which Spirits are conveyed from it by the Nerves into the several Parts of the Body for the performing of the Animal

> actions or motions; for all voluntary motions are performed by the help of these Spirits.

Not bad, eh? "Spirits" aside, this is a reasonably accurate general description of the nervous system. Even the spelling is leaps and bounds ahead of Markham. However, he was mistaken regarding the adrenal glands, which he charmingly referred to as "Deputy-kidneys":

> Over the Kidneys a little more outward, and about an inch from them, there stand two Kernels, which are known by several Names, from the several uses that Authors have ascribed to them. Some call them Deputy-kidneys, because of some resemblance they have to the true ones in their frame, and because they have been thought to assist them in separating the Urine. Bartholine calls them Black-choler Cases, from an Opinion that they receive black Choler from the Spleen. Others have impos'd other Names on them, which we shall not recite. It is not long since they were first found out. They are in an Horse about as big as a Garden-bean.

The true role of the adrenals in secreting hormones involved in the regulation of the metabolism wasn't sorted out until the 19th century. Incidentally, you may have noted the reference to "Black-choler," meaning black bile. This was still a time when Hippocrates's humoral theory of health held sway. Even after this daffy theory was finally abandoned, Snape's anatomy remained a standard reference until close to modern times.

Now we're about to enter the 18th century, and things are going to change. Finally. To this point veterinary medicine has been a potpourri of theories and remedies largely based on superstition, tradition, and anecdotal observation. As we've seen, although some of these weren't too unreasonable, the majority did more harm than good. But now something we might recognize as objective science is beginning to emerge, albeit in halting little steps. There are several places we could choose to launch this part of the story, but let's begin with Pope Clement XI's cows.

It's the year 1713, and a physician at the Santo Spirito hospital in Rome has been summoned to the Vatican.

CHAPTER NINE

Lancisi

"The Holy Father? Is he unwell?" Lancisi asked. He wiped blood from his hands onto his apron.

"No, he is very well. God be thanked," the messenger said. He handed Lancisi a small envelope sealed with red wax. The messenger, a young boy no older than 12, stared straight ahead, obviously taking care to avoid looking at what the doctor had been working on.

Lancisi nodded and took the envelope. He waved his hand to dismiss the young man before opening the letter.

> Dr. Giovanni Maria Lancisi, you are required to attend his Holiness, Pope Clement XI, immediately to discuss a matter of greatest urgency.

He's not sick, but it's urgent, Lancisi thought. He had been personal physician to the popes for more than 20 years, since Innocent XI, and in the early years, urgent summons had always been for flare-ups of gout or sudden fevers or, in Innocent's case, terrible pains from kidney stones.

But more recently he had been consulted on matters of public health, such as the relationship between malaria and the swamps of the Tiber. But these were rarely urgent questions. Unless it had to with the mass dying of cattle, rumours of which were moving quickly through the city.

Lancisi tucked the note away. He glanced at the table of human hearts beside him, in various stages of dissection, and sighed. Hopefully Clement would be quick. It was going to be very hot today, and these would spoil.

It was a short walk to the Vatican from the hospital. The guards recognized the pope's doctor and admitted him immediately. Although Lancisi had been there often enough that he could find his way blindfolded, a page guided him through the ornately decorated maze of halls leading to Clement's office.

Clement sat behind an enormous desk studded with candlesticks, inkwells, blotters, seals, boxes, and books. Stacks and stacks of books abounded, some teetering precariously. Innocent had been more orderly, but also more severe. Clement had a long mournful face, unlike the pinched Innocent, or the jowly Alexander. He looked downcast and tired. Most remarkably, grey stubble covered his chin and cheeks like on a beggar. His predecessors had all had fashionable goatees and moustaches, and Clement was typically clean-shaven.

Lancisi stepped towards the side of the desk, preparing to kiss the pope's ring, but Clement shook his head and waved him back. It was a weary gesture, not an impatient one. He motioned him to sit in a red leather chair opposite the desk.

"Lancisi, thank you for coming so quickly."

"Of course, Your Holiness. I hope my visit finds you in good health and that the urgent matter is one of general consultation rather than personal exigency." Lancisi bowed his head briefly after speaking and smiled at the pope. They had always been on friendly terms.

"You are correct, Doctor. Thanks be to the saints and to our saviour, my body does not trouble me today. But my mind is sorely vexed."

"Your mind?"

"My mind and, truth be told, my soul." Clement paused and then leaned forward. "Lancisi, cattle are dying. Great numbers of cattle. I have here reports from Spoleto, Urbino, Rimini, Ancona, Bologna, Ferrara . . ." He lifted a sheaf of papers in the air, evidently the letters from across the Papal States. "Not just great numbers of cattle, Lancisi, but in some places, *all* the cattle."

"The cattle plague?"

"Yes. That is what they tell me."

"Many people will starve."

"This is what weighs on my soul, Lancisi. You are the most learned man I know. This is why I called for you. But you are a physician for humans, not for animals, so I hope this is not a violation of some oath or moral principle I am unaware of."

Lancisi shrugged and said, "The great Bernardino Ramazzini, my colleague in Padua, gave a speech in Venice two years ago when the cattle plague was a scourge there. There were murmurings that it was beneath a renowned physician to consider the maladies of beasts, to which he said, 'I know that someone will accuse me of having deviated to veterinary medicine; let them bark as much as they want.'"

Clement smiled. "Happily, I am not one to bark. And any who does, I shall bring them to heel quickly."

"It is all one medicine."

Clement raised an eyebrow.

"A cow's heart looks identical to man's heart in all aspects, only it is larger."

"Physically larger. Not metaphorically," Clement said.

"You are correct." Lancisi bowed his head slightly. "I speak, of course, purely of the physical. The metaphorical, the spiritual, the eternal, I leave to you and your cardinals, your bishops, and all the saints."

Clement nodded and cleared his throat. "And this Doctor Ramazzini, did he propose a remedy to the doge?"

"After a fashion, yes. Ramazzini had made a study of pox among corpse-bearers. He noticed that when many became ill at the same time, it was among those who had shared touch. They had been very close to one another, not just in the same village. And close to the bodies of those who had died of the same disease. Others in those villages without this closeness remained well."

Clement furrowed his brow. "Spread by touch, not by bad air, vapours, miasma?"

"Yes. Even the ancients knew it was possible for some sicknesses to spread this way, but this knowledge is never applied. People ask for remedies, for salves and draughts. The Venetians were using quinine, which is a fine medicine and useful to lessen the fever of cattle plague. Ramazzini did not think it would stop the spread."

"If no medicine, what then?"

Lancisi began to count off on his fingers. "First, he told them to bury the dead cattle as deeply as possible. Second, clean their barns and stalls well. Third, isolate the sick and those in contact with the sick. Fourth, do not mix the herds. Fifth, feed the best, freshest hay available."

The pope tented his fingers on the desk in front of him and nodded slowly. "And this stopped the cattle plague?"

"After a time, yes, it appeared to have."

"After a time? And this is what you suggest I order as well?"

Lancisi waited a moment before answering. He had been considering this problem since speaking to Ramazzini. The Venetian cattle plague outbreak had been small, and the measures taken only worked slowly. The situation here in the Papal States was a hundred-fold worse, and the pope, as lord of these lands, depended more on the health of the herds than the doge of Venice did. Ramazzini had started down the right path, but there was further to go. Much further.

"Holy Father, if you'll indulge me, this is what I believe needs to be done. All the ill cattle must be slaughtered. We cannot just wait for them to die and in the meantime spread the plague. Once slaughtered, they are to be kept whole and buried and covered in lime, so that the virulence cannot escape."

"Slaughter all of them, and not make use of any of the meat? People will be unhappy."

"They will be unhappier when there are no cattle left alive anywhere. It is said that this was the plague that caused the great famines towards the end of the Roman Empire."

"I am to be like poor Romulus Augustulus?"* Clement smiled and drummed his fingertips together.

"No, Your Holiness. He had Odoacer, the barbarian, to contend with."

"Of course. And we will stop this plague. That is all? Slaughter and bury under lime?"

"No. Also, healthy cattle must not be moved, and herds must be kept separate — as Ramazzini did in Venice. And the meat of well cattle should be inspected before sale or consumption. But especially important is that you enforce penalties for those who do not follow the slaughter edict, especially those who waste time by attempting cures instead."

"Penalties? What have you got in mind, Doctor?"

"The severest. Scofflaws should be threatened with hanging, drawing, and quartering. And even your priests, if they should offer to pray for the healing of stricken cattle, as some are wont to do in the smaller villages, these priests should be sent to the galleys."

Clement leaned back and closed his eyes.

For a moment Lancisi thought he might have even fallen asleep, which would put him in an awkward position. But soon there was a sharp intake of breath, and Clement opened his eyes again.

"I trust you, Lancisi. You have never given me bad advice. I will have the orders drawn up immediately."

Like most popes before the 20th century, Clement XI was of noble birth. In an age when almost everyone was at least somewhat religious, you only had to be somewhat religious to ascend to the papacy. Your connections and your political skills were often more important than your piety, for the pope was not only the head of the Catholic church, but also secular lord over the Papal States, a sovereign territory that occupied a large portion

* Last emperor of the Western Roman Empire, deposed in 476. Magnificently, and ironically, named after both the mythical founder of Rome and the first Roman emperor.

of central Italy. Consequently, the fates of cows were more important to him than you might have otherwise guessed.

The dreaded cattle plague that struck Clement's herds is known now by its German name, "Rinderpest," which just means "cattle plague." Rinderpest was an ancient disease, known already to the Egyptians and likely one of the ten plagues allegedly visited on them by the God of Israel in the Book of Exodus. It had mysteriously come and gone at various times since then, but this time it was not going — it was just continuing to spread. Rinderpest is extremely virulent and able to quickly kill astonishingly large numbers of animals. Concerns were mounting about the impact on the food supply. Famine was always just one drought or plague away. And famine meant unrest, and unrest could spiral in a way that any head of state, whether a pope, a prince, or a king, would take pains to avoid.

Lancisi's advice marks the first sustained advance in the history of veterinary medicine. His advice was successful, and catastrophe was averted. As he noted in his conversation with Clement, isolation of sick animals was not a new concept. What was new was its rigorous enforcement over a large area, not just with the occasional herd, like the sheep pox in FitzHerbert's time. And what was especially important was that the news spread, and the method became widely known and followed. A year later rinderpest crossed the Channel to England. There, Thomas Bates, King George I's personal surgeon, took charge of the control efforts using what he had learned about Lancisi's methods. The only difference was that Dr. Bates did not suggest the drawing and quartering of those who refused to slaughter their sick cattle. Instead, he proposed strict financial penalties. Happily, this appears to have been just as effective, and England was free of rinderpest within three months. Less happily, on reoccurrences later in the century, slaughter edicts were more haphazardly enforced, and rinderpest frequently raged out of control. An epizootic (animal epidemic) in 1865 in the UK resulted in an especially disastrous outbreak, with rail and steamship transportation of infected cattle resulting in more rapid spread.

In the late 18th century, the Dutch, Germans, and Danes experimented with vaccination against rinderpest. They still had to contend with superstition as local pastors sometimes proclaimed that rinderpest was the judgment of God upon the sinful farmers, but gradual progress was made, much in the fashion of "two steps forward, one step back."

Lancisi's principles of enforced slaughter and isolation remained sporadically in use, and then, two-and-a-half centuries after his time, most rigorously and enthusiastically by the Soviet Union and communist China. Disobedient farmers were merely imprisoned, though, not drawn and quartered.

Ultimately a combination of slaughter and mass vaccination led to rinderpest being declared eradicated worldwide in 2011.* Lancisi's edicts are the earliest thread in veterinary science that we can follow through to today.

Giovanni Lancisi's contribution to what we now call veterinary epidemiology was seminal, but he was not a veterinarian and did not think of himself in those terms. He was first and foremost interested in human health. Animal health was only important insomuch as it provided useful comparisons, and for how it impacted the well-being of people by destroying an important part of their food supply.

So far, among our protagonists, only Ubar and Oswald were fully dedicated to animal health as a profession, and they are fictional composites. Nobody who treated animals full-time left enough of a mark on history to be named, short of Shalihotra, about whom we know little, and the likely mythical Palakapya. FitzHerbert and Vegetius were polymaths for whom animal health was just one interest among many. Fébus was a nobleman who happened to have a passion for hunting dogs. The paleolithic old woman was probably closer to Lancisi than any of these — a healer of people who applied her knowledge and skills to animals when it was important to do so.

This is about to change. We are about to leave the leeches, folk healers, dilettantish noblemen, and dabbling physicians behind. With one exception,

* The last known case was recorded in Mauritania in 2003.

all the remaining stories will focus on professional veterinarians, more or less in the modern sense. In other words, people specifically trained in and dedicated to the diagnosis and treatment of animal diseases. Interestingly, rinderpest has a further role to play in this transition.

First the exception, though. It involves another healer who saw more similarities than differences between the maladies of humans and animals. Around the same time that Dr. Giovanni Lancisi and Pope Clement XI were conferring about the cattle plague, far across the Atlantic and most of the way across the wide plains of North America a young woman was thinking about horses.

CHAPTER TEN

Sinopa

This was her favourite spot, on the ridge above the village, facing east to where the sun rose beyond the ocean of grass, her back to the mountains. She closed her eyes and took a long, deep breath. She would have until the sun was two finger breadths above the horizon before her sister, Iinisskimmaakii, came up here and bothered her. Iinisskimmaakii always slept late. Sometimes this was a nuisance, but today it was a blessing. Sinopa enjoyed the peace. Often, she would see animals at this hour. Some would come quite close. Ground squirrels, deer, jackrabbits, little voles. Sometimes even her namesake, the kit fox. She knew how to be very still.

Her father said that she would have made a good hunter if she had been born a boy, but instead she was becoming a healer, a medicine woman. But it was useful for a medicine woman to know how to be still too, especially when she was treating horses, which is what Sinopa really wanted to do. She wanted this more than anything in the world. Her aunt, Saoyi'kitstaki, said that she could become the most important medicine woman among the Siksika. But that was many years away still. For now, she was still learning. Today there would be a test. Apisi's horse was peeing blood. This was painful for the horse, and it made him slower. Apisi was not only Sinopa's brother, but also one of the most important hunters in the tribe. Buffalo hunt season was coming soon. It was important that Apisi's horse be in the best health possible.

"What are you thinking about?" came a voice from behind her. Sinopa jumped. Iinisskimmaakii was early.

"Sorry, sister!" Iinisskimmaakii giggled. "I didn't mean to scare you."

"Yes, you did!"

"Okay, I did. But it was funny, wasn't it?"

"A little bit. But only a little bit." Sinopa tussled her sister's hair and gave her a gentle shove on the shoulder when she sat down beside her.

"So, what were you thinking about?"

"Apisi's horse."

"You are going to make him better?"

"I am going to try. I listened to Ksista'paaki carefully, but listening and doing are two different things."

"Do you use the same medicine you use for people?" the younger girl asked as she absent-mindedly plucked blades of grass from the ground in front of her. Sinopa smiled to herself, thinking that although her sister was close to 15 years old, she often still behaved like a little girl.

"Of course we do. Horses are people too. We are the Siksika. There is where the Piikani live," she said, pointing to the south. Turning to point to the east, she went on, "And there, the Kainai." She paused to smile at her sister and then pointed into the valley below her where a small herd of horses grazed along a stream beside the village, "And *there* is the Ponokáómitaa. The Horse Tribe. Some mingle with the human tribes, and some live on their own."

Iinisskimmaakii laughed. "They came from the south, right, sister?"

"Yes, in the time of our grandmother's mother. Do you know the story?"

"I do! But tell it to me anyway."

Sinopa swept her long black hair out of her face and took a deep breath before beginning. "Many years ago," she said, "an orphan named Long Arrow was adopted by an Elder. When he became a man, Long Arrow asked the Elder what he could do to repay him. The Elder told him about a lake, a long way away, where spirit people lived at the bottom who had mysterious animals that could run like elk but carry things like dogs. 'So we call them elk dogs,' the Elder said. 'Bring these elk dogs back to our people, Long Arrow. They will be useful.'"

Sinopa paused and glanced at her sister. She was listening, rapt. So much like a small child, Sinopa thought again. "So, Long Arrow went far way until he eventually found the lake. There he met a young spirit who told Long Arrow to swim to the bottom and find an old spirit man wearing a long robe —"

"Long Arrow could hold his breath so long?" Iinisskimmaakii interrupted.

Sinopa laughed. "Hush. It's a story, sister. Strange things happen to story people. So, the young spirit told Long Arrow to watch the bottom of the old spirit man's robe. If he saw the old spirit man's feet, he would be given an elk dog. So he waited. One day. Two days. Three days. On the fourth day, the old spirit man's robe caught on a stick, and Long Arrow could see that his feet were hooves!"

Iinisskimmaakii clapped and laughed. "Long Arrow was patient! He got to see the hooves because he was patient."

"Yes, you're cleverer than you look, sister. Patience is the lesson in this story. And we must always be patient with animals. So, the old spirit man was the master of the elk dogs. He gave Long Arrow his robe, which would cause the elk dogs to stay with the people; and a beautiful colourful belt, which contained all the elk dog songs; and a whole herd of elk dogs. Elk dogs are horses. This was the Horse Tribe. We Siksika have been allied with them ever since."

"Wonderful, sister!"

"So, they are like people. And like people, I will use for them what Apistotoke, the Creator, has caused to grow on the land for our use."

"There is a plant that will stop blood from being in the pee?"

"There is. And you can help me find it. I have some, but we need more. It is a tall yellow flower, with petals that hang down —"

"Aókspiiyipisátssaisski!"

"Yes! That's the one. I'm impressed again. You were paying attention when I pointed out different plants."

"Sometimes." Iinisskimmaakii grinned and stood up. "Let's start looking! Apisi will be angry if you are slow in making his favourite buffalo-hunting horse better. He loves that horse more than he loves either of us!"

Sinopa, and her sister and the rest of them, are of course fictional. But the use of healing traditions in horses among Indigenous Peoples in North America is not. The Peoples of the Plains, such as the Siksika (a Blackfoot tribe in what is now Alberta), the Só'taeo'o Cheyenne, and the Nakota Sioux, among others, placed enormous value on their horses. As the story illustrates, this value was practical and emotional, uniting the motivations for veterinary medicine.

The mouthful-of-a-name yellow flower is the upright prairie coneflower (*Ratibida columnifera*). The Sioux used it for treating urinary infections in horses, so I allowed myself the extrapolation that the Blackfoot likely did as well. While I could not find any modern scientific references to this specific indication for prairie coneflower, it is known to contain pharmacologically active compounds and has proven efficacy in treating fevers. This was only one example from an astonishingly broad plant-based pharmacopoeia available to healers such as Sinopa.

This is not to say that Siksika medicine women were any more adept or wise than their counterparts elsewhere. Sinopa was practising a type of medicine that was common worldwide in hunter-gatherer societies that kept domesticated animals.* Analogous remedies can be found in many traditions, across vast expanses of time and space. Many of these remedies were useful, at least to some extent, but some were not and were as dubious as Oswald's ritual with the bloated cow. For example, the Sioux recommended pulverizing the roots of field

* Traditional Maasai medicines for cattle and Mongol treatments for horses, being just two examples.

sagewort (*Artemesia campestris*) into a paste and putting that on the face of a sleeping man to make it easier to steal his horses. Applying this paste must have been a delicate procedure, lest you awaken the man and cause him to become suspicious of your intentions regarding his horses. And somehow, burning the roots of a specific type of yucca was supposed to make it easier to catch horses. The catching, stealing, and retaining of horses was obviously of great importance to the First Peoples of the Great Plains.

These traditions died out with the colonization of the continent by Europeans and the subsequent violent suppression of Indigenous culture and knowledge. Consequently, they do not play a role in the development of the modern veterinary profession, but I tell the story to highlight the fact that the desire to care for animals is a universal human trait, independent of culture and history. In spirit and style, Sinopa's story is more akin to the first four in this book, but in the timeline it belongs here. The golden age of Indigenous horse culture was from about 1700* until the apocalyptic collapse of the buffalo herds around 1880.

From the grasslands of what is now Southern Alberta, we now shift back to Europe and take up the story of rinderpest again, this time in France.

* Horses were present in the Americas up to the Pleistocene but went extinct about ten thousand years ago. Hernán Cortés brought them back in 1519 when he arrived on the continent in Mexico. By the mid-to-late-1600s, the Comanche people, in what is now the American Southern Plains, were the first to use larger numbers of horses. This radically transformed their lives and culture. It allowed for more productive buffalo hunts, much more mobility to find better places to live, and greater success in warfare. One can only imagine the reaction of their enemies, the first time they saw Comanche warriors surging over the ridge astride these great alien beasts.

CHAPTER ELEVEN

Bourgelat

With a slight squeeze of his legs, Claude urged Theo from a walk into an amble. The big bay stallion did this so smoothly that to the casual observer nothing had changed, only that they were a little faster now. But a horseman would see that one of Theo's hooves was always off the ground, whereas before, when he was walking, there were alternatively one or two off the ground. Claude was especially proud of teaching Theo to amble. It had not been easy. The stallion had been considered unbreakable by the cavalry, but to Claude he was just another equine puzzle to solve.

The rolling hills outside of Lyon were ideal riding country, especially in the early morning hours before the mid-summer heat rose. The air was perfectly balanced between warm and cool. Mist still clung to the valley bottom. The air had the sharp scent of cedar and rosemary. The only sounds were the songs of nightingales and thrushes and the rhythmic thud of Theo's hooves on the hard, dry earth as they approached the hilltop.

Claude glanced at the sun, now cresting the wooded ridge on the far side of the valley. Beams of yellow light shot out from between the trees.

Like the fingers of God, Claude thought, smiling to himself. If God had fingers, which struck him as unlikely. The overnight post would be delivered soon, so he should head back down. Today was Thursday. Bertin was to have met the king on Tuesday, August 4. The post usually took a day and a half in good weather. Claude wasn't sure if the meeting was in the morning or the afternoon. If it was the morning, Claude would have his answer today. Otherwise, tomorrow.

He squeezed his legs and made an almost inaudible sound to convey his instructions to Theo. The horse sped up to a trot and then a canter. Claude wouldn't gallop him today. The ground was uneven coming down into the valley. Besides, excessive haste was unseemly for a gentleman and the canter was Theo's best gait. Rider and horse were as one. Fluid and quick, like water coursing down the hillside.

"The post? Has it come?" Claude asked the stable hand as he dismounted.

"Yes, Monsieur Bourgelat. Not ten minutes ago."

Claude stroked the horse's neck and said, "Well done, Theo. Well done."

The boy took the reins from Claude and led Theo away.

"Give him an extra carrot," Claude called after them. He crossed the yard and entered the house.

The morning's mail lay in a neat stack on his desk in the dark wood panelled study at the back of the house. The large leaded glass windows showcased his small vineyard out back. Normally Claude would have changed out of his riding clothes, made himself comfortable in his chair, and then rang for a cup of coffee, a glass of wine, or a flute of champagne, depending on the day, before addressing himself to the post. Today he stood and rifled through the pile quickly. We'll soon see whether it's a champagne or coffee day, he thought. There were letters pertaining to his various appointments with the Academy of Sciences in Paris and the Berlin Academy, as well as the usual queries from equestrians and trainers from all over Europe respecting his *Elemens d'hippiatrique*. A masterpiece, if he did say so himself. The definitive guide to horsemanship. And then there was a letter from Denis Diderot. Claude had been contributing articles for Diderot's great encyclopedia. This was important, but not as important as what he was looking for.

There it was, the last one. Of course it was the last one. A letter sealed with the large red insignia of the Royal Council in Versailles. A decree by King Louis XV himself. Or a note of polite rejection from his secretary.

Now Claude sat down and wiped his hands, which had begun to sweat, on his pants. He turned the letter over and then tore it open.

It was the decree.

Claude let out a little yip of triumph and pleasure. He quickly scanned the elaborately penned document, all flourishes and swooshes, to confirm the key point. Yes, the king would provide the funds to transform Claude's Academy of Equitation in Lyon into a school dedicated to the teaching of animal diseases and their treatments. The world's first veterinary school. Claude grinned to himself and rang the brass bell in the corner of his desk. It was a champagne day.

He read the decree more carefully now, especially the statement

> . . . to open a school where would be taught publicly the principles and method to cure the diseases of livestock that would procure imperceptibly for the agriculture of the realm the power to conserve livestock in times where this epidemic desolates the countryside.

"This epidemic" was rinderpest. While objectively a terrible thing, this dark cloud over French food production had produced a silver lining for him. His friend, Henri Leonard Jean Baptiste Bertin, was controller general of finances. This was the only way Claude had gotten the king's ear in the first place. Merely being the renowned author of *Elemens d'hippiatrique* was not enough. And Bertin had made it clear that the only way his dream of a school for equine medicine would be funded from the royal purse would be if it also addressed the cattle plague and other animal diseases of economic importance to France.

"This is the best plan," Bertin had said as they talked over brandy at the controller general's sumptuous home in Paris. "This way Louis wins. The farmers win. And you win."

"And the horses," Claude added.

"And the horses!" Bertin laughed. "But mostly Louis!"

"Long live the king!" They had both said.

The champagne arrived on a small silver tray. Gaston, the footman, had been in Claude's employ for many years.

"Good news, sir?"

"The best news, Gaston. The very best news." He raised his glass towards the old man and beamed. "Tell Luc I'll be riding Theo again this evening. It will be a long ride."

Although veterinary medicine can be said to have many parents, plus grandparents and aunts and uncles, it is Claude Bourgelat who is most often cited as being the "Father of Veterinary Medicine." He was the right person, at the right place, at the right time to light the spark that led to an explosion of veterinary schools. In 1760 there were zero. Within the decade, there was a second school in France, and ones in Italy, Austria, and Germany. Hungary, Sweden, and Denmark followed soon after. Today there are at least 539 veterinary colleges and universities worldwide.*

Two forces transformed veterinary medicine from a shambolic folk art into an Enlightenment Era science. These forces were central to Louis XV's charge to Bourgelat. The first was that the population of France was growing rapidly, and the food supply was tenuous. Famine was always just a bad harvest or, more pointedly, a surge of rinderpest away. Famine meant unrest. And unrest meant . . . well, in 32 years' time, Louis XV's grandson would find out exactly what it meant. It was ever more critical that the food supply be secure. Developing a veterinary science for the problem of epizootics was a critical part of that security.

The second was that France had been more or less continuously at war for the entire century, beginning with the War of Spanish Succession in

* Wikipedia lists 539 but, curiously, several countries are missing, including Sri Lanka, Colombia, and, notably, Russia, so the real total is likely closer to 600.

1701 through to the Seven Years' War, referred to by some historians as the *real* "First World War" because it ranged across three continents. The Seven Years' War was in its fifth year when the veterinary school in Lyon was approved. Cavalry was at its apogee as the principal instrument of military power. As we saw in the introduction, horses were being killed and wounded in numbers that kept military planners awake at night. Properly trained veterinary surgeons could at least help return the wounded horses to service faster. For Louis XV, this argument was not enough on its own without the need to control rinderpest, but it was a factor. As we will see, the military factor had a greater role to play in the founding of the first British veterinary college in 1791.

That covers right time and right place, but right person? Veterinarians could have done worse than Claude Bourgelat as "father" of their profession. Not only was he smart enough to see the opportunity and exploit it properly, but also he appears to have been a genuinely interesting person. He was born in 1712 in Lyon. His father was the local sheriff, who was also a silk trader who, for some unknown reason, sported the interesting nickname "The Golden Feather." Claude went to a Jesuit school where he studied law, but he had the reputation of being a bit of a rake and man about town, more interested in carousing than studying. Nonetheless he qualified as a lawyer and practised for a short time. There is a story that he left law after being unable to cope with the moral quandary of having to defend a man he knew to be guilty. It seems, however, that his career change was due as much to pull as push. Horses were his passion from a young age, and he felt the draw of the equestrian world strongly. He joined the cavalry and quickly established himself as someone with a special touch and intuition who could train the most challenging horses. Bourgelat was also able to quickly synthesize all the available knowledge about the species and then innovate from that base. It is said that Frederick the Great, king of Prussia and one of the greatest military strategists of all time, consulted with Bourgelat regarding the optimal gait to use during a cavalry charge.

When Bourgelat became convinced of the need to professionalize what passed for veterinary medicine at the time, he threw himself into the research of equine anatomy, pairing it with a study of current human medicine. In this he was helped by two Lyonnais doctors. This combination allowed him to lay the foundations for what would be taught at L'École vétérinaire de Lyon.

Bourgelat was also a child of the Enlightenment. His project was very much in the spirit of those times where everything was thrown open to question and inquiry. Reason was replacing superstition. Science was replacing theology. His contemporaries included Benjamin Franklin, Voltaire, Immanuel Kant, Jean-Jacques Rousseau, Adam Smith, Carl Linnaeus, David Hume, and the list goes on. You'd be hard pressed to find another era with a greater concentration of original thinkers. Also, as mentioned, Bourgelat was involved in Denis Diderot's encyclopedia project, contributing almost two hundred articles. This was the world's first true encyclopedia. The first volumes came out in 1751, preceding the storied *Encyclopaedia Britannica* by 17 years (I'll touch on the *Britannica*'s interesting take on veterinary medicine in the next chapter). Diderot's encyclopedia, gathering all objectively verifiable knowledge about the world, is often considered the central project of the Enlightenment.

An intellectual environment that fostered rational and objective thought was, of course, key to permitting the foundation of a veterinary school. But what in retrospect seems obvious and inevitable was anything but at the time. Consider, for example, that the world's first dental school didn't open until 1840.*

Yet, as groundbreaking as L'École vétérinaire de Lyon was, it's also important not to overstate its sophistication. The curriculum was limited to anatomy, physiology, pharmacology, and general medicine, plus splinting and bandaging. The only entrance qualifications were proof of baptism and good conduct. The youngest student was an astonishing 11 years old, and the oldest were in their thirties. Bourgelat discouraged anyone from applying who also had an interest in human medicine because he feared they would inevitably give up on veterinary medicine because of the poor pay relative to that offered physicians. Some things never change. In another distant

* The Baltimore College of Dental Surgery, in the USA.

echo of today, Bourgelat strove to strike a balance between rote memorization of the course material and hands-on practice for the students. This balance is still the subject of endless debate in veterinary colleges today. For the hands-on component in Lyon, the students even maintained an herbarium, where they grew some of the medicinal plants discussed in their pharmacology course.

The Lyon school was a success. So much so, that Bourgelat set his sights higher and lobbied for a second school at Alfort, in Paris. This was granted, to the chagrin of his rival, Philippe-Étienne Lafosse, who had been petitioning to be allowed to start his own veterinary school. Lafosse was likely the better veterinarian, but he was also the unluckier one. The one without connections in the court. He went on to write the seminal *Cours d'hippiatrique, ou traité complet de la médicine des chevaux* (*Veterinary Studies, or Complete Treatise on Equine Medicine*) in 1772, and he is sometimes referred to as the "Father of Veterinary Anatomy," which is a definitive grade or two below Father of Veterinary Medicine. But for Bourgelat's friendship with Bertin, this chapter would have had a different title. However, it worked out well for the profession. Bourgelat was still the right man.

And the timing was not only right, but also highly fortuitous. Two years later, in 1763, France was hit by another major rinderpest outbreak. This time Bourgelat's students, freshly minted "veterinarians," were on the case. The plague was brought under control, and famine was averted. It's impossible to know with certainty now how much the new veterinary education can be credited with this success, perhaps there were other forces at play, but regardless, Louis XV was convinced. The original grant to open the school had been temporary, but the king was so pleased by the control of rinderpest that in 1764 he decreed that the school be funded permanently and henceforth called the Royal Veterinary School.* Bourgelat continued to direct and teach at both schools — in Lyon and Alfort — until his death in 1779 at the age of 67.

* Now the ENVL, École nationale vétérinaire de Lyon, albeit in much bigger and more modern digs.

Another great centre of Enlightenment thought was across the Channel in Scotland. There we meet a man every bit Claude Bourgelat's equal in skill and the desire to modernize veterinary education but who, like the hapless Lafosse, lacked the right connections.

CHAPTER TWELVE

Clark

The tall man in the white periwig and long red velvet coat spoke first. "It is the poll evil, is it not, Mr. Clark?"

"Aye, it is, Lord Maitland, I'm sorry to say. And a bad case of it too." Clark, a shorter, dark haired man in a tan coloured coat, carefully felt a swelling on the horse's neck. He spoke to the horse quietly.

The two men were standing in a dimly lit stable. It was raining steadily outside. Maitland called to a stable boy to bring another lantern.

"You've had Mr. Sinclair here," Clark said in a flat tone, trying to conceal his annoyance.

"I have. He's done good work with my horses before. But this confounded him."

"Of course, it did," Clark muttered. "He incised it up here." He pointed to a ragged wound on the top of the swelling and then waited for the other man to step forward and look before going on. "It's far too high. And too superficial. And he did it too early. You must wait for it to ripen, to soften. And if all that were not enough, he applied some foul tar-based unguent or some such." Clark rubbed his finger beside the wound to collect a little black substance and sniffed it. He made a face and offered his finger to Lord Maitland.

Maitland took a step back and held his palm up. "Thank you, Mr. Clark. I believe you. But you can help her? She is an excellent coach horse. I would so hate to lose her."

"Yes, that I certainly can, sir. If your boy will fetch another lantern and then hold her steady, I will demonstrate the correct way to treat poll evil."

"I would be in your debt, Mr. Clark. All in the district say you are the best farrier. I should have called you first."

Clark shrugged and knelt down to examine the contents of the leather case he had brought with him. After a moment he stood up again and held out what looked like a long metal needle with a chord attached to it. "This is a seton, sir."

Maitland motioned for the stable boy to raise the lantern, and he peered carefully at the object. "A seton?"

"Yes, by means of this sharp needle I will insert the chord into the swelling from its *lowest* point. It is ripe enough now. See the fluctuance?" Clark prodded the bottom edge of the swelling to demonstrate that it was soft.

"What purpose does the chord serve?"

"It will be left in place to draw the foulness out. When the swelling has abated, I will return and remove the chord."

"Will it not fall out?"

"That is a very good question, sir." Clark looked at Maitland, recalibrating his opinion of the man a little. "In some locations on the body I will use a thread or a wooden button to secure the chord, but here, up on the withers, it is not necessary. Feed her well, treat her well, and do not allow her to become bored."

Maitland nodded. "Very well, Mr. Clark. But I do not wish to watch. When Sinclair cut into her, she became quite agitated. I do not need to see that again. I have a soft heart for my beasts."

Clark smiled. Really not so bad for a lord after all, he thought. "As you wish, sir. But let me assure you that she will react very differently with my treatment. In fact, she is unlikely to react at all. The tiniest flinch, perhaps.

You see, this needle is exceedingly sharp, much sharper than Mr. Sinclair's knife, I'll warrant. And I am quick and will insert it at a point where she already has pain, so she will hardly notice this prick. And after she will feel relief as the foulness leaves her and the pressure subsides."

"Thank you, Mr. Clark. I hope you are correct."

James Clark was correct. The "poll evil" — which, in case you haven't figured it out, was a large abscess above the horse's shoulder — was successfully treated using the seton type drain. The Scottish farrier recounts this specific case in his 1788 book, *A Treatise on the Prevention of Diseases Incidental to Horses from Bad Management in Regard to Stables, Food, Water, and Exercise. To which are Subjoined Observations on some of the Surgical and Medical Branches of Farriery*. Book titles were often long in the 18th century. Going forward we'll refer to it as Clark's *Treatise*.

Clark was quite prominent in his field at the time, but, probably because of his working-class background, we don't know when or where he was born or even when or where he died. He was also a member of the Odiham Agricultural Society, which would be instrumental in founding the first veterinary college in the United Kingdom, in London in 1791. We'll get to the Odiham story soon. Clark wrote two other books, with the *First Lines of Veterinary Physiology and Pathology* intended to become a textbook for a second British veterinary school in Edinburgh. He was promised a professorship at the planned Edinburgh school, so he turned down an offer to be appointed head of the London school when that position suddenly came open in 1793. But the promised Edinburgh professorship fell through. For convoluted political reasons, the School of Veterinary Studies in Edinburgh was not started until 1823. Clark did not have anything like the connections or social standing that Bourgelat did.

But for our story he is important. One aspect of his writings that puts him well above his peers is his focus on the welfare of the horses. Bourgelat appears to have been kind to horses too, using less harsh and punitive training methods than many at the time, but James Clark goes a step further. The following passage from Clark's *Treatise* makes that clear:

> The reason why horses are generally rendered vicious is maltreatment of some kind or other on their being first handled, and in the breaking; the effects of which remain longer with some than with others, according to their tempers or dispositions; for they possess these peculiarities as well as mankind, of which there are considerable variety, and which are not necessary here to particularize, as the method to be observed with them in general is the same. For if we wish to have them docile and tractable, we will succeed much better by familiarity and caresses and gentle usage, than force and chastisement. We ought, in this respect, to take a lesson from the Arabians, whom I had occasion formerly to quote, and whose horses, it is to be observed, are remarkably tractable and docile, and from whom the finest horses in Europe are descended. Buffon* says in vol. III page 368, "That as the Arabs live in tents, these tents serve likewise for stables. The mare and her foal, the husband and his wife and children sleep together promiscuously.** The infants often lie on the body, or on the neck of the mare or foal, without receiving any injury from these animals . . . The Arabs never beat their mares, but treat them gently, and talk and reason with them. They are so careful of them as to allow them always to walk, and never spur them, unless the occasion be very urgent."

Clark goes on for several pages to discuss how important it is to be gentle and kind when treating horses. He warns the reader that horses have an excellent sense of hearing and are very intelligent, so they will

---

* Georges-Louis Leclerc, Comte de Buffon, an 18th-century French naturalist.

** He only meant sleeping in full view of one another.

remember someone who has mistreated them and will recognize them by voice alone. This horse will consequently become more and more difficult to handle by that individual. Clark is also clear that even the tone of voice the person uses is important and warns several times against "chastisement." It's a remarkable insight when you consider that the wisdom of wielding the carrot rather than the stick is apparently still foreign to some people today when it comes to dealing with behavioural issues in animals, and even in people.

The other element of note is his admiration for the Arabs. This is a beautiful example of the opening of minds that occurred during the Enlightenment. Make no mistake, most Europeans still considered themselves to be superior in most respects to peoples elsewhere in the world, but at least some were finally willing to try to learn from them.

So, what of the fact that Clark was a farrier? A split was coming soon, but in the late 18th century, farriery and veterinary medicine were synonymous in Britain, at least when it came to the treatment of horses, which was the primary animal-health concern. The third volume, charmingly called "Volume the Third," of the *Encyclopaedia Britannica*,* which covered the letters M through Z, published in 1771, does not have an entry for "veterinary" or "veterinarian." It hops right from "veteran" to "viales," the latter being the minor Roman Gods who guard the roads and, evidently, are of greater significance than "veterinarian." Recall that Thomas Browne already used the word "veterinarian" back in 1646. In "Volume the Second" (C to L), however, there is a detailed 39-page section on farriery. The introduction is illuminating with respect to the opinion of the learned classes regarding the state of equine health care:

* Indulge me while I share the full title of the encyclopedia, every bit as florid as Clark's book, and with abundant eccentric capitalization: *Encyclopaedia Britannica; or a DICTIONARY OF ARTS and SCIENCES, COMPILED UPON A NEW PLAN. IN WHICH The different SCIENCES and ARTS are digested into different Treatises or Systems; AND The various TECHNICAL TERMS, &c. are explained as they occur in the order of the Alphabet.* Whew.

> Farriery, the art of curing the diseases of horses. The practice of this useful art has been hitherto almost entirely confined to a set of men who are totally ignorant of anatomy, and the general principles of medicine. It is not therefore surprising, that their prescriptions should be equally absurd as the reasons they give for administering them. It cannot be expected that farriers, who are almost universally illiterate men, should make any real progress in their profession. They prescribe draughts, they rowel,* cauterize etc. without being able to give any other reason for their practice, but because their fathers did so before them. How can such men deduce the cause of disease from its symptoms, or form a rational method of cure when they are equally ignorant of the causes of diseases and the operation of medicines. The miserable state is this useful art.

Okay, Mr. Macfarquhar and the Society of Gentlemen in Scotland, tell us what you really think about farriery. The article goes on to apologize for the paucity of good information but promises that they have selected from "the best authors" (uncredited) "such a system of practice as seemed to be formed on rational principles." What follows is a mix of herbs, potions, clysters, ointments, purges, and poultices, largely unchanged since Gaston Fébus's time. Plus, of course, the requisite advice regarding bleeding, "especially when their eyes look heavy, dull, red, and inflamed; as also, when they feel hotter than usual, and mangle their hay." The recommended treatments of wounds and abscesses comes closer to the rational principles they claim to espouse, but still fall short of what the farrier, James Clark, was doing at around the same time. Incidentally, Clark was not illiterate. In fact, he had read the works of the ancients, including our old friend Vegetius.

So, here we see the class divide at work. Farriers were not gentlemen. To leave this "useful art" in the hands of the lower classes was no longer deemed appropriate. The time had come to elevate farriery to a gentleman's

* The spiked disk on a spur.

pursuit. This brings us then to the Odiham Agricultural Society, founded in England in 1783 by a group of concerned "gentlemen of rank, fortune, and ingenuity," in addition to a few "intelligent farmers." In 1785, a Thomas Burgess, one of the founders of the society, moved:

> That Farriery is a most useful science and intimately connected with the Interests of Agriculture; that it is in a very imperfect neglected state and highly deserving the attention of all friends of Agricultural economy.
>
> That Farriery, as it is commonly practised, is conducted without principle or science and greatly to the injury to the noblest and most useful of our animals.
>
> That the improvement of Farriery established on a study of the Anatomy, diseases and cure of cattle, particularly Horses, Cows and Sheep, will be an essential benefit to Agriculture and will greatly improve some of the most important branches of national commerce, such as Wool and Leather.

As mentioned, as an indication of his renown and perhaps his distance from common farriery practice, James Clark was also permitted to be a member of the society. One presumes he was rolled into the "intelligent farmer" category.

The English agriculturalist and travel writer Arthur Young was also a member. He was a witness to the French Revolution and was in Paris during the storming of the Bastille in 1789. He also visited the veterinary school in Alfort. Young reported that it educated "over one hundred pupils from different parts of France as well as pupils from every country in Europe except England, a strange exception considering how grossly ignorant our farriers are." Again, the disdain of farriers, presumably couched in "present company excepted" terms when he discussed this at Odiham Agricultural Society meetings where Clark was present. Young's intelligence on the French veterinary school prompted the society to send a couple of students there (one of whom, William Moorcroft, more than qualifies for a chapter of his own). They soon came to the perhaps obvious realization that sending people to France was not a viable long-term solution. The French Revolution further underlined this. Therefore, with the help of a French veterinarian, Charles Benoit Vial de Saint Bel, the society formed a London Committee in 1790 with the aim of opening the first British veterinary school there.

Vial de Saint Bel, originally Vial, of Saint Bel, and later Vial Sainbel, was an interesting character. He had been a professor at the school in Alfort but moved to England to pursue an interest in the growing field of thoroughbred horse breeding. Here he came to the attention of the wealthy Irish gambler Dennis O'Kelly. O'Kelly's partner was, incidentally, the leading London brothel madam of the age. An interesting individual in any context, but in the relevant context, especially interesting because O'Kelly also owned Eclipse, one of the most celebrated racehorses of all time.

Eclipse was a marvel. It's worth taking a moment to digress into his story. Medical histories, including this one, I'm ashamed to say, spend almost all their time focused on the doctors, with the patients given minor walk-on parts at best. Eclipse deserves to take centre stage for a paragraph, at the very least.

Eclipse was named for the 1754 solar eclipse during which he was born. He was chestnut with a narrow white blaze on his face and a white right hind leg. He's described as not having been especially attractive, especially with his rump higher than his withers, and he was large for the time, very large, standing just over 16 hands (64 inches or 1.63 metres) at a time when the average thoroughbred was 14 hands. Eclipse was spirited and wilful enough that some thought was given to gelding him, but when this exuberance was channelled into racing, this notion was set aside. Eclipse was undefeated as a racehorse, including in 11 King's Plate races. Not only was he undefeated, but also his victories were often with astonishingly wide margins. At the time, a common expression for dominance in any field was "Eclipse first, the rest nowhere." Everyone in England in the 1770s and '80s knew who Eclipse was. After retiring from racing, Eclipse became a prodigious stud, commanding a 50-guinea fee.* It is estimated that 95 percent of all thoroughbreds have Eclipse as an ancestor. Secretariat, Northern Dancer, and Phar Lap were direct descendants.

And then he got colic and died on February 26, 1789.

It was a major news event and a tragedy that gripped the nation. O'Kelly asked Vial de Saint Bel to perform the autopsy. The result was his book, *Elements of the Veterinary Art, Containing an Essay on the Proportions of the Celebrated Eclipse*. A key finding was that Eclipse had an unusually large heart, weighing 14 pounds versus the usual eight. At least some of his athletic prowess was attributed to this. Much like Bourgelat's skill in synthesizing human medicine and equine anatomy was for the Lyon school, Vial de Saint Bel's involvement as a college-trained veterinarian in Eclipse's case (although, it must be said, entirely post-mortem and thus not especially helpfully from the poor horse's perspective) was a catalyst in the promotion of formal veterinary education in Britain. People paid attention, and it therefore became relatively easy for the Odiham Agricultural Society to raise the necessary funds. Under the patronage of the Duke of Northumberland, "The Veterinary College, London" was founded on February 18, 1791. This was the beginning of the end of "farrier" as a synonym for "veterinarian." Farriers would continue to legally practise some limited form of veterinary medicine in Britain for another 150 years, but over that period college-educated veterinarians increasingly dominated the field.

A macabre footnote to Eclipse's story is that his hooves were preserved as inkstands. In that era this was considered a fitting tribute to a beloved celebrity. Much in the manner of medieval saints' relics, there were at least five "genuine" Eclipse hoof inkstands. (Insert joke about the real reason he was so fast.) One of them was gilded and given to King George III for use as a goblet. George is now believed to have lived with bipolar disorder,

* Understanding historical prices is always challenging, but the average labourer earned around 12 pounds a year in 1800. This is about 11.4 guineas because in the quirky old British monetary system a pound was 20 shillings while a guinea was 21. So, 50 guineas was a lot of money. At the other end of the wealth spectrum, it may be of interest to note that in *Pride and Prejudice*, Mr. Darcy is described as being very rich, and his income was ten thousand pounds a year (or 9,524 guineas).

so it's not outside the realm of possibility that he may actually have drunk from this during his periods of mania. Eclipse's skeleton is still on display at the Royal Veterinary College, as it is now known, in London.

Before we leave the world of James Clark, Vial de Saint Bel, and Eclipse, it's interesting to note the parallels between farriery and human dentistry. Class divisions played a role in the development of the dental profession as well. Tooth pullers, often barbers, were traditionally looked down upon by professional gentleman physicians as mere technicians. Physicians had the same attitude towards the barber surgeons, but the latter ultimately manoeuvred themselves into acceptance in a larger physician-surgeon profession, leaving the barber dentists behind to form their own separate profession. This is an oversimplification of why humans have one set of doctors to look after their teeth and another to look after every other part of their body (and their mind — even psychiatrists have to be MDs). But the basic premise is true — tooth pulling was felt to be beneath the dignity of 18th- and 19th-century physicians. It was the realm of working-class, mechanically inclined individuals with grubby fingers and unfashionable accents. Had veterinarians been as prissy, we could have a whole separate profession today dealing with the diseases of horses' feet. Doctors of Farriery. Instead, now farriery is to veterinary medicine what pedorthics* is to human medicine.

* Human feet have a confusing welter of job titles for the people catering to them. In addition to pedorthists, there are also podiatrists, chiropodists, orthotists, and orthopaedists. Feel free to pick one of these appellations instead for this analogy if it makes more sense to you.

CHAPTER THIRTEEN

Erxleben

Although there was still enough ink in his quill, Johann absent-mindedly dipped it into the pewter inkwell while he paused to consider his next words. The sound of garbled singing rose from the square below his open window. Students, he thought, and smiled to himself. Drinking and singing again. At least they're not drinking and fighting.

He leaned forward and closed the window.

Then it came to him, and he began to write:

> It may be said that those people to whom we entrust our sick animals to be cured have gradually got to know various medicines through experience, which they found capable of curing this or that disease. Fine, but it happens all too often that those medicines that have a good effect on a certain disease at one time, on another time, when the animal is suffering from something that appears to be similar, have the completely opposite effect and cause the greatest

> damage if you use them with full confidence simply because of their previously noted good effect. If one is not able to investigate the causes that brought about the disease, it will only depend on blind luck and mere guesswork as to whether the medicine chosen blindly kills the animal or at least increases the disease, or whether, to the contrary, it will bring them back to health. And above all, these so famous aids which long experience is said have proven to be reliable, are unfortunately all too often harmful, if not outright criminal, and not infrequently superstitious, and the use of which one must be very careful to avoid.

He waved the sheet lightly to dry it and then read it over. Was it too much? No, it was probably not enough. His *Theoretical Lectures in Veterinary Medicine* were going to be nothing if not forthright and bold. Students did not come to hear the credulous nonsense of centuries past regurgitated but only the newest and the most scientific. His position and, in fact, income as professor here at the Georg-August University in Göttingen depended on it. The students paid per lecture. But he would do this for free if he could afford to. All over Europe, science was ascendant, and he, Johann Christian Polycarp Erxleben,* would do his part, one way or another.

Erxleben stood from his chair and began to pace. His apartment was small, and by no means luxurious, but it was no matter to him. That the Freiherr von Münchhausen** had responded positively to his proposal and allowed him to start a scientific veterinary school was enough. Five paces with his long legs, and Erxleben was at the other end of the room.

---

* In case you're curious, as odd as it sounds to modern ears, Polycarp was a common name at one time in religious families. You might think it has something to do with fish, but it's Greek for "many fruit." Saint Polycarp was an important early Christian martyr. This becomes vaguely relevant soon.

** For all you Terry Gilliam fans out there, yes, this is the eponymous Baron von Munchausen. The film was based on a 1785 book of stories about Münchhausen. The old baron, in addition to being a pioneering patron of veterinary education, was also a renowned storyteller, exaggerating and emphasizing wit and whimsy when recounting anecdotes from his long life and many travels.

He went back and forth a few dozen times, careful to step lightly lest the landlord below complain of the noise.

More than lectures was needed. He had been to the school in Lyon and observed how the students were taught there. Although the course material prepared by Monsieur Bourgelat was impressive, the practical element was less so. Yes, they dissected animals, grew medicinal herbs, mixed various ointments, and learned to fashion splints and bandages, but they observed very few examinations and treatments of actual patients. Erxleben would do this differently. But how? How would he even find suitable patients in the city? How would a farmer out in the valley of the Leine or the foothills of the Harz know to bring his beast to the university?

Erxleben sat again and poured himself a glass of wine from the decanter on his desk. He was not a habitual drinker, but sometimes he found it helped him to calm down and focus. He sipped the wine and thought of his father, who had been a deacon in Quedlinburg and who was quite fond of wine, although in moderation and with a connoisseur's taste that their means kept in check.

That was it. Priests and parsimony.

Erxleben was struck by two thoughts. The first was that even in these modern times, deacons and priests were often the most influential people in their communities. He would spread the word among them. They would tell their parishioners that scientific treatment for their sick animals was available in Göttingen. His second thought was that as an added incentive he would provide the treatment at no cost. It would be part of the university curriculum, so any expenses associated would be covered by the students' fees. He did not think the Freiherr would object to this unorthodox arrangement. Although Münchhausen was 50 years older than him, Erxleben found him to be much more progressive and open-minded than many in even his own generation.

Erxleben briefly considered a second glass to celebrate his exciting idea but decided against it. Instead, he would channel this energy into writing a few more pages. The square outside was quiet again. The only sounds now were the scratching of his quill on paper. He loved that sound.

Johann Christian Polycarp Erxleben is, after Bourgelat and Clark, the third of our Enlightenment Era veterinary heroes. We're spending so much time in the late 18th century because it was pivotal. And also because these are fascinating people. Take Johann. In 1770 he founded the first veterinary college in Germany at the remarkably young age of 27, and he was yet another of these polymaths, with broad interests and an astonishing amount of energy. This sort of thing ran in the family. The story above references his father, but it's his mother whom he took after. Dorothea Christiane Erxleben was the first woman to study medicine in Germany (to be clear, human medicine). This was a time when women needed special permission from the government to go to university at all. She began medical school at the University of Halle in 1741, and the next year wrote a book titled *A Thorough Inquiry into the Causes Preventing the Female Sex from Studying*. Various family issues forced her to postpone finishing her degree. The family needed money, however, so in 1747 she began to practise. At that time, it was quite common to do so without a degree. Local physicians protested this and sued her for quackery, but ultimately Frederick the Great himself, King of Prussia, ruled that she should be permitted to practise so long as she passed an examination and submitted a dissertation to the university. She did this with no trouble and was granted an official MD degree on June 12, 1754. This is a hundred years before the first women were allowed to practise medicine in the US and Britain; however, she was, sadly, the singular exception in Germany.

But back to her son. What Erxleben brings to our story is a quickening emphasis on scientific principles paired with hands-on practice for the students. This is the model veterinary schools follow today. Although, it must be said, treatment at modern teaching hospitals is not typically free, nor are the clergy involved in drumming up patients. He also expanded the scope of veterinary education from Bourgelat's focus on the care of cavalry horses and epizootics in cattle. The species were the same, but the number of diseases and conditions considered were considerably broader. Erxleben's veterinary school in Göttingen was a significant step forward.

Incidentally, the quote in the story above is my translation from Erxleben's book. I mention that to illustrate a limitation inherent in writing this sort of history — not everything has been translated into English. In fact, very little relevant to veterinary history has. Claude Bourgelat's and Gaston Fébus's French-language works are the prominent exceptions. Erxleben's contemporaries included outstanding veterinarians writing in Italian, Danish, and Swedish, but I can read German, not Italian, Danish, or Swedish, so I chose to feature him. It's worth keeping in mind that this factor will make it seem like the bulk of the action was in the English-speaking world, but in fact advances were being made in many countries. It's just so much easier to access the English language information in North America. And, for me, to some extent, the German. So, with that segue, I'm going to jump ahead to a story with a personal angle set in 1832 in Leipzig, Germany. It involves canary medicine.

My grandfather Herbert Schott was a lawyer. He always hoped that I would follow in his footsteps, albeit thousands of kilometres away in Canada, where my parents and I had immigrated to when I was a baby. It was thus with surprise and no small measure of disappointment that he greeted the news that I had decided to pursue veterinary medicine (more or less accidentally, as it happens, but that's the subject of a previous book). One of my cousins was also showing interest in veterinary medicine, so poor Herbert was overheard to plaintively wonder where this sudden interest in "agriculture" came from in the Schott family. Our ancestors had been academics, theologians, artists, and millers. To his credit, he adjusted and even became supportive. The most tangible evidence of the latter was when a packet arrived from him after I had been in practice for a year or two. It was a veterinary book. Not just any veterinary book, but an 1832 first edition of the *Vollständiges Rezeptbuch für Tierärzte, Landwirte, so wie überhaupt für Eigentümer von Haustiere*

jeder Art (*Complete Pharmacopoeia for Veterinarians, Farmers, as well as for Owners of Pets of All Kinds*).

To be honest, although as a book lover I was delighted to have such an old volume in my hands (it's still easily the oldest original book I own), I didn't spend much time looking through it. It was in a difficult-to-read Gothic typeface on yellowed paper and was daunting in size at 734 pages. There was no way I was going to actually read this. Cool, but kind of useless. In flipping through, it seemed to mostly consist of old-timey drug recipes for horses (tincture of this, dram of that). I was in my twenties then and had other things on my mind. It's only now, 30 years later, writing this book that I realize what a treasure Opa gave me. I wish I knew more about how this book came about, but I can find no references to it anywhere. The editor and compiler is a D.A.P. Wilhelmi, who has left no other traces of himself whatsoever, and the publisher has been out of business for 150 years. What is remarkable about it is the "for owners of pets of all kinds" in the title. It is one of the first times (the other is the focus of the chapter after next) that I know of in almost five hundred years since Fébus's *Livre de chasse* that anything substantial related to what we now call companion animal medicine appears in print anywhere, and the *Livre de chasse* is very focused on hunting dogs and only happens to mention medical care as part of a comprehensive look at their husbandry. But here, in Wilhelmi's pharmacopoeia, we're not just talking about hunting dogs but animals of all sorts, including pet dogs and cats. These even appear in the subtitle: *Effective Medicinal Formulas from the Practices of the Best and Most Experienced Veterinarians for All Inner and Outer Illnesses of Horses, Cattle, Sheep, Goats, Pigs, Dogs, Cats, and also Poultry*. This is the first reference anywhere to the medical care of cats. And, yes, although not mentioned on the cover (unless I've mistranslated "poultry"), canaries* are featured

* The history of canaries as pets is interesting. Sensibly enough, they originate in the Canary Islands, off the coast of Morocco. They began to be sold in Europe in the 16th century, but the traders were canny, only selling males to prevent competition from local breeders. They are difficult to sex, though, so eventually some females were accidentally sold as well, and nature took its course. By the 19th century, the biggest centre of the canary trade in Europe was in the Harz Mountains, not that far from Leipzig, so perhaps that's part of the reason Wilhelmi gives more pages to them than to the presumably more common geese and ducks.

too. In fact, they get a whole chapter to themselves. Here are the first few canary diseases the book discusses:

1) The Fever — With this it is said that it is necessary to immediately seek to find out the causes of it and then to use the necessary tools. You can tell by the sadness of the birds that they are sick. When they have the fever and the little bump on the back of their tail is swollen, they woo weakly, remain silent, and only occasionally utter a few plaintive tones. Then the bump just mentioned must be cut away from them, but with as much care as possible, because they very easily die from deep wounds.

2) Gout — These little animals can also be very easily attacked by this; but, as experience has shown, the easiest way to lift it is to rub warm wood pulp into their thighs. Canaries' gout is a kind of tearing inside their thighs and legs, which is accompanied by trembling of one or the other part of the same, sometimes also with some swelling.

3) The Epilepsy — These little birds are afflicted by the said evil in such a way that they fall down as if dead and flap all around themselves with their wings. In this case, the instillation of the purest Dels* (perhaps the Provencal Dels, or very pure and still young linseed oil) has often proved very salutary.

4) The Moult Disease — In the autumn every year the canaries moult and lose their entire plumage. At that

---

* I have no idea what specifically this is, but it appears to be an oil of some sort. Interestingly, only very recently have we started using medium chain triglycerides (a type of oil) for epilepsy in dogs. It's the active ingredient in a special diet I give to my own epileptic sheltie. It has dramatically reduced his need for medication. Go figure. Linseed oil is, however, an omega-3 fatty acid, not a medium chain triglyceride, and is not known to have any anti-epileptic properties, although there are other health benefits associated with it.

> time, they must be kept in the room so that, while their bodies are naked, they are not endangered by the cold or the draught, which would result in their early death. They should also not be given any cooling food at this time, but especially no lettuce.

Especially no lettuce, eh? The reference to "cooling food" shows that the writer was still under the sway of the long dominant humoral theory of medicine, wherein anything taken into the body, be it food, liquid, or medicine, is categorized as being either cooling or heating (regardless of serving temperature), and also either moistening or drying (also regardless of how wet or dry it actually is). Lettuce would be considered cool and moist, and thus inappropriate for a naked shivering little bird. We obviously don't subscribe to this theory anymore, but we still discourage the feeding of pale lettuces, such as head and iceberg, to birds. Humoral medicine was not conclusively discredited until the 1850s.

Wilhelmi goes on to give advice regarding eight more canary ailments, including diarrhea, constipation, foot disorders, catarrh (excessive mucous buildup), and something called "Windsucht" in German, which directly translate as "wind addiction." This odd name makes even less sense when you read the description: "Here the bird blows up part or all of the skin." And the treatment is "Make a small opening in the skin with a pin so the air can escape." Ah. So, this ends up being subcutaneous emphysema, a condition where there is an injury or disease in part of the respiratory system, especially the air sacs, which causes air to leak into the space under the skin. The bird does indeed "blow up part or all of his skin." The treatment is the same now. We poke a hole, although we use something fancier and more sterile than a pin. I am tempted to say that one also needs to hold the bird very tightly to prevent it from careening off like a punctured balloon, but that would be silly. So, I won't say it.

All this canary medicine from two hundred years ago is curious and amusing, but it also represents a milestone. With very few exceptions, the history of Western veterinary medicine has thus far been one entirely given to practical considerations and rationales. You will recall that at the beginning of this book I detailed the two principal reasons for people to want to provide medical care for animals: practical and emotional. Practical reasons would still be overwhelmingly the most common factor for another century, but change was beginning. The emotionally motivated veterinary medicine many of us practise today had its origin around the time people began deflating "wind-addicted" canaries.

Why was that? I don't mean specifically to question why people would begin to puncture swollen canaries but, more generally, why they would begin to care for animals who served no practical purpose.

There were several forces at work. The first was the Industrial Revolution, which from the mid-18th century onwards fuelled a rapid expansion of the middle class. Going back to ancient times, the aristocracy had enjoyed the company of pets. Now much larger numbers of people had the means to copy aspects of the fashions and lifestyles of the upper classes, and so it was with companion animals. We can see this phenomenon at work today in the Global South, where rising levels of prosperity often lead to increases in pet ownership and increased demands for modern veterinary services. Pets were, and are, a luxury and a status symbol in economically ascendant households. Pay attention to movies set in the Regency Era, such as the ones based on Jane Austen novels, and watch for the dogs and the caged little birds. Contrast this to movies set in the Middle Ages.

The second factor was the shift from Enlightenment Era values of rigorous rationality to the more emotion- and feeling-centred approach to life favoured by the Romantic Period, beginning around 1800. Look, for example, at the work of George Stubbs, the English painter famous for his depictions of horses, dogs, bulls, and various wildlife, with human-like expressions. This is the same phenomenon that I touched on in the first chapter when I described the change in how horses were painted in battle scenes. Or consider the poetry of Samuel Taylor Coleridge or William Blake and their frequent use of anthropomorphized animal subjects. Johann Wolfgang von Goethe said, more prosaically and directly: "He who tortures animals is unsouled and missing God's good spirit, however noble he may look, he should never be trusted."

A well-known example of the tenor of the times can be found in Robbie Burns's 1785 ode "To a Mouse," which begins:

> Wee, sleeket, cowran, tim'rous beastie,
> O, what a panic's in thy breastie!
> Thou need na start awa sae hasty,
>   Wi' bickerin brattle!*
>
> I wad be laith** to rin an' chase thee
>   Wi' murd'ring pattle!***
>
> I'm truly sorry Man's dominion
> Has broken Nature's social union,
> An' justifies that ill opinion,
>   Which makes thee startle,
> At me, thy poor, earth-born companion,
>   An' fellow-mortal!

Burns bridged the Enlightenment and Romantic periods, but this poem is pure Romanticism in its brilliant expression of the newly emerging empathy with "fellow-mortal" creatures. Not that such empathy didn't exist before — we know it did — but never before had it been in the forefront of public discourse.

Another example is the public reaction to the 1760 dog cull in London. There was an outcry when a rabies panic prompted an officially sanctioned attempt to kill every dog in London, often by quite brutal means. Rabies was, to be sure, a significant public health threat, but this cull was viewed by many people as a barbarous overreaction, not befitting a civilized society. There were many voices in favour as well, but that there was any debate

* A clattering sound.

** Loath, reluctant.

*** A spade-like tool used to clean the plough-shears.

at all was new and yet another milestone in the redefinition of Western civilization's relationship with animals.

This open expression of ideas that previously would have been considered frivolous or irrational or, to go back a little further still, sacrilegious, came at a time when the economy was expanding and more people could afford to look after animals kept simply as companions. Fortuitously, it also came at a time when a newly emerging profession was positioning itself to meet that need.

The medical treatment of pets was still a quiet undercurrent, though. The dominant thrust of veterinary medicine remained practical throughout the 19th century, focused on horses and farm animals.

A painting by Ernest Crofts of the 52nd Light Infantry capturing a French artillery battery at the Battle of Waterloo on June 18, 1815. Note the eyes of the chestnut horse in the foreground.

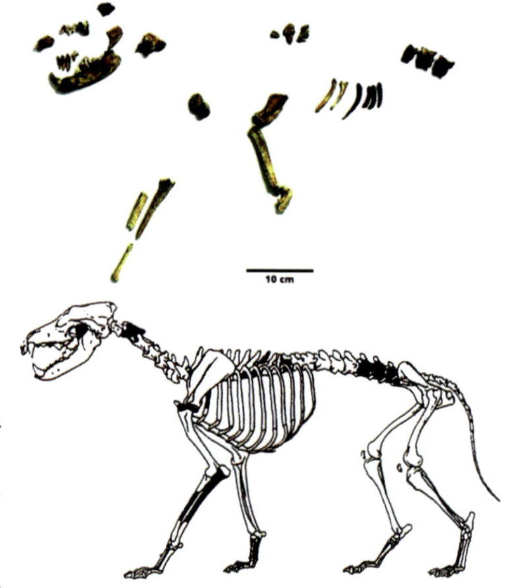

Luc Janssens et al, University of Leide

The remains of the Bonn-Oberkassel dog, c. 12,205 BCE. It's astonishing how much of a story a few scattered bones and teeth can tell.

Wikimedia Commons

Prince Rupert's dog Boye. Mid-17th-century painting attributed to Louise, Princess Palatine. Confusing a fluffy white poodle with Satan seems ludicrous until he is brought to the vet for a nail trim.

Historic UK, Wikimedia Commons

A Parliamentarian soldier shoots Boye at Marsden Moor while Prince Rupert looks on, aghast. Rupert is made to appear absurdly popish, and poor Boye is far more devilish looking here than in Louise's sympathetic portrait.

Evenki (Tungus) shaman in Siberia, 18th century. Those look like elk antlers, not reindeer, but the artist was going on descriptions and was presumably not a biologist.

Wikimedia Commons

Wellcome Collection

The clay seal of Urlugaledinna, late third millennium BCE. I believe the famous tongs are the parallel pair of slightly curved pointy objects left of the man, hanging from the tree. We use similar instruments today (assuming these are not depicted to scale relative to the man).

Rama, Wikimedia Commons

Wooden model of a cow giving birth, Egypt, c. 1990–1786 BCE. Everyone — cow, calf, and man — looks remarkably relaxed.

Mary Harrsch, Wikimedia Commons

Stone relief showing a cavalryman pulling an arrow out of his horse. Tang Dynasty (618–907 CE), China.

Nebamun, a court official, and his cat hunting birds on the Nile, c. 1350 BCE. Imagine an alternative history where this continued and spread, so that the modern camo-clad hunter went into the field with his prized tabby instead of a bouncy black Lab.

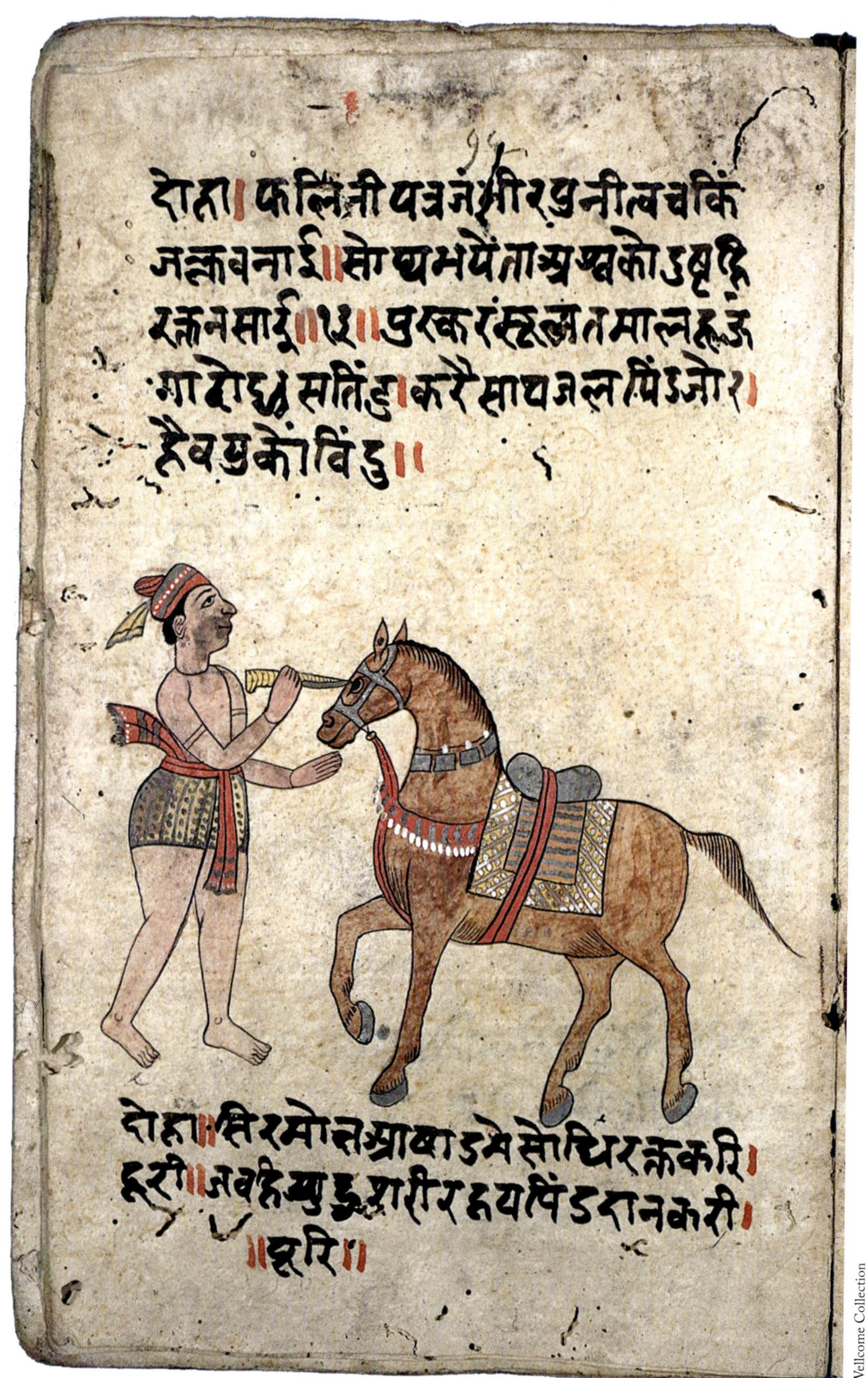

An 18th-century edition of the Indian third-century BCE *Shalihotra Samhita*, depicting what looks like a remarkably bold eye operation, given the large, sharp object and the evidently fully conscious horse. File under "Don't try this at home."

A Byzantine Encyclopaedia of Horse Medicine (McCabe, 2007)

A folio from the Hippiatrica, a fifth- or sixth-century CE Byzantine collection of Ancient Greek writings on the medical care of horses. The upper picture shows a horse being drenched (i.e., being given a large volume of medicated fluid orally) to induce diarrhea, perhaps to clear parasites or an obstruction. The lower picture shows the horse's delight at the successful completion of this procedure.

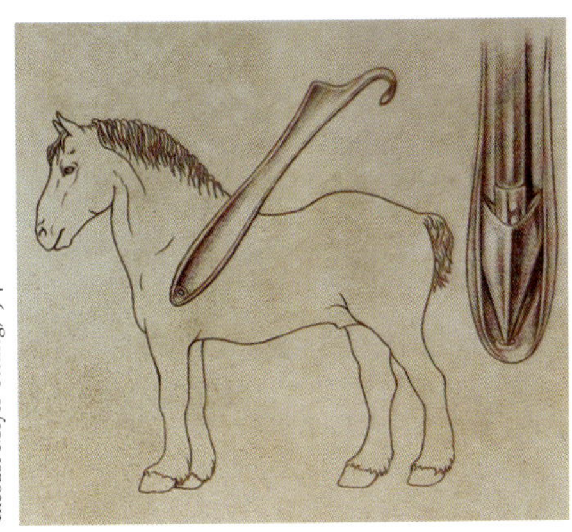

Theodor Meyer-Steineg, 1914

The Greek spoon of Diocles, used to remove arrows. (The spoon is enlarged relative to the horse for the illustration. Obviously.)

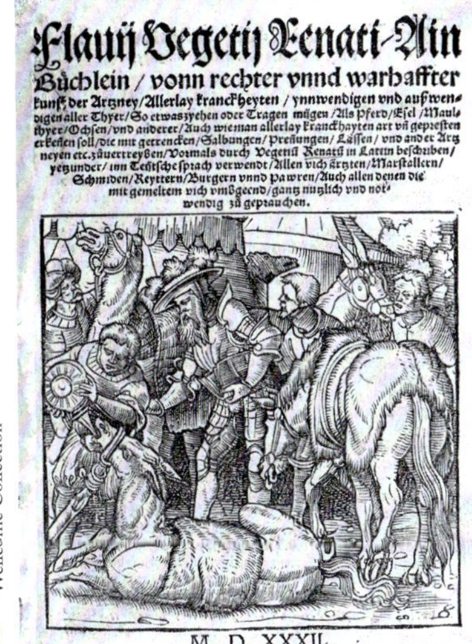

Flauij Vegetij Renati-Ain
Bůchlein / vonn rechter vnnd warhaffter
kunst/ der Artzney/Allerlay kranckheyten / ynnwendigen vnd außwen-
digen aller Thyer/So etwas zyehen oder Tragen mügen /Als Pferd/Esel /Maul-
thyer/Ochsen/vnd anderer/Auch wie man allerlay kranckhayten art vñ gepresten
erkennen soll/die mit getrencken/Salbungen/Preßungen/ Lässen/ vnd ander Artz-
neyen etc. zůuertreyben/Vormals durch Vegetiũ Renatũ in Latein beschriben/
yetzunder/ inn Teütsche sprach verwendt/Allen vich ärtzten/Marstallern/
Schmiden/Reyttern/Burgern vnnd pawren/Auch allen denen die
mit gemeltem vich vmbgeend/gantz nutzlich vnd not-
wendig zů gepr̄auchen.

M. D. XXXII.

Wellcome Collection

Cover of a 1532 German edition of Pelagonius's fourth century CE *Ars veterinaria*. There's a lot going on here, but drenching of the unfortunate horse seems to be the focus again. It's unclear whether depicting everyone in 16th-century styles was a way to make the thousand-year-old book look topical, or whether it was due to ignorance of how Romans actually dressed.

Morgan Library

A 1469 Italian manuscript of Columella's first century CE *De re rustica*, depicting a horse being bled. The veterinarian appears to be wearing chaps without pants on underneath. A fashion which I am relieved to report has not caught on in the profession.

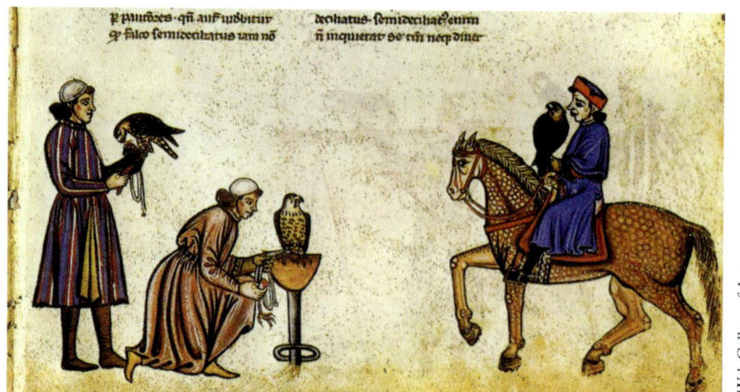

Web Gallery of Art

ABOVE: A page from Emperor Frederick II's *De arte venandi cum avibus*. I'm not entirely sure what's going on here, but it appears to demonstrate the various ways you can hold a falcon.

Juan Alvarez, c. 1390, Libro de Menescalcia y Albeyteria

RIGHT: Mending a fractured metatarsal in the Spanish translation of a 1390 Moorish veterinary text, *Libro de menescalcia y albeyteria*. It is interesting to note the harness suspended from a beam, allowing the horse to take its weight off the injured leg. There's not much happening here that would be out of place today.

Veterinary Medicine: An Illustrated History (Mosby, 1996)

A 1209 Arab manuscript, *Kitāb al-bayṭara* (*Book on Veterinary Medicine*), showing a sick zebu cow being helped to lie on a mat of liquorice root. This begs the question, "Why liquorice root specifically?" There have been studies to show that it can relieve skin inflammation, so maybe the cow was itchy.

Livre de chasse, Gaston Phébus, fol. 40v

A vivid illustration from Gaston Fébus's 1389 *Livre de chasse* (*Book of the Hunt*). If you look carefully, you will see eight different treatments being applied. Remarkably for the time, none of them look bizarre.

Husbandrie.

shyp. And the housbandes holde an opynion, that it shall the rather cease. And whan the beaste is slaine, there as the murren dothe appere betwene the fleshe and the skynne, it wyll ryde uppe lyke a ielly and frothe an inche depe or more. And this is the remedy for the murren, Take a smalle curteyne corde, and bynde it harde aboute the beastes necke, and that wyll cause the bloudde to come in to the necke, and on eyther syde of the necke there is a vayne that a man maie fele with his fynger: and than take a bloud yren, and set it streight uppon the vayne, and smyte hym bloudde on bothe sydes, and let hym blede the mountenaunce of a pynte or nyghe it, and than take awaye the corde, and it wyll staunche bledyng. And thus serue all thy cattell, that be in that close or pasture, and there shal no mo be sicke by goddes leue.

Longe sought, and remedy therefore.

There is an other maner of syckenesse among bestes, and it is called longe sought, and that sicknes wil endure long, and ye shal perceiue it by his hostynge, he wyl stand muche, & eate but a littel, and waxe very holowe and

Passage on a "syckenesse in bestes . . . called longe sought" from a 1548 printing of Anthony FitzHerbert's *Boke of Husbandry*. With "longe sought," the poor cow will cough 20 times a day and will gradually waste away, so it is some sort of chronic respiratory disease. FitzHerbert's suggested remedy is to cut open the dewlap and fill it with a poultice of "feitergrasse." He concludes this passage with, "Thus I have seen used, and men have thought it hath done good."

Reeman Dansie, lot 687

MARKHAMS
Maister-Peece.
Containing all knowledge belonging to Smith, Farrier, or Horſe-leech, touching the curing of all diſeaſes in Horſes.
Deuided into two bookes.
The firſt, containing all cures Phyſicall.
The ſecond, all belonging to Chyrurgery.
The Sixth Impreſsion, corrected and enlarged by the Author.
Geruaſe Markham.

LONDON
Printed by Iohn Okes. 1644

Circulating Now, National Library of Medicine

The cover of a 1644 printing of *Markham's Maister-Peece* with a considerably more modest title than the 1610 original. The illustrations mostly depict reasonable-looking actions, although what's going on in panel ten at the bottom is anyone's guess.

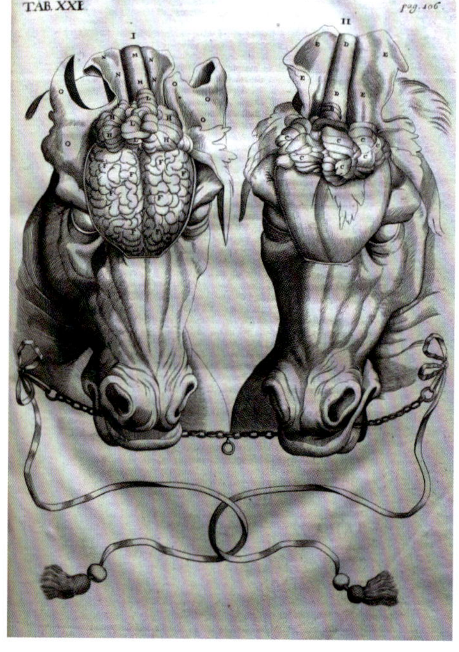

Edward Worth Library

One of 49 copperplates in Andrew Snape's 1683 *Anatomy of an Horse*, depicting a remarkably graphic dissection of the brain. This must just have been for completeness and intellectual curiosity, as there would have been no need for anyone to go poking around there in a live horse (they do not get the brain cysts seen in cattle and sheep).

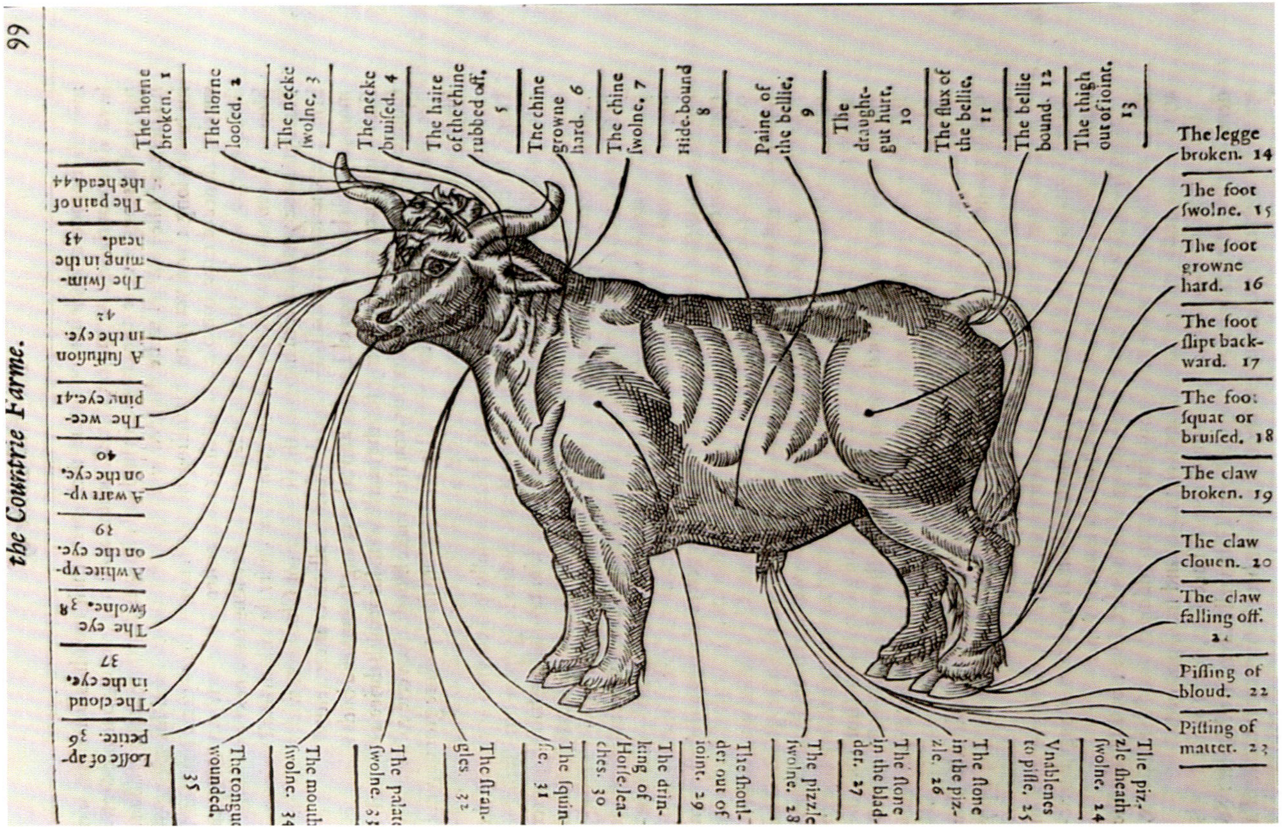

A disease chart published in Gervase Markham's *The Countrey Farme* (London, 1616). He copied and translated it from Charles Estienne's *L'agriculture, et maison rustique*. The 44 marked conditions are interesting to look through, starting with "The horne broken" and then running clockwise around the body, through eight different conditions involving pissing or "the pizzle," and ultimately circling back to the top with "The pain of the head."

The Science Museum, Science & Society Picture Library

The Royal Collection, Windsor Castle

L'Empirique by Edmond Tschaggeny in 1845, depicting what is likely a cow-leech swindling peasants. The facial expressions say it all. The peasants are worried but cautiously hopeful, and the leech is arrogant and judgmental. The dog looks suspicious, and the cow just looks defeated.

Wellcome Collection

Giovanni Maria Lancisi (1654–1720), c. 1728.

Rijksmuseum, Wikimedia Commons

A 1745 illustration by Jan Smit entitled, *God's Striking Hand over the Netherlands, through the Plague in Cattle.* Why God would do such thing and what the cattle had done to deserve this fate is not specified. If the humans have sinned, why not strike them directly? However, the nattily dressed gentleman in the bottom right appears to have a plan. What it might be is also not specified.

Provincial Archives of Alberta, Paul Coze fonds, PR2006.0508/10

The Nakoda Nation was adjacent to the Siksika and was another of the great horse cultures. This photo was taken of a camp near Banff, Alberta, in 1930.

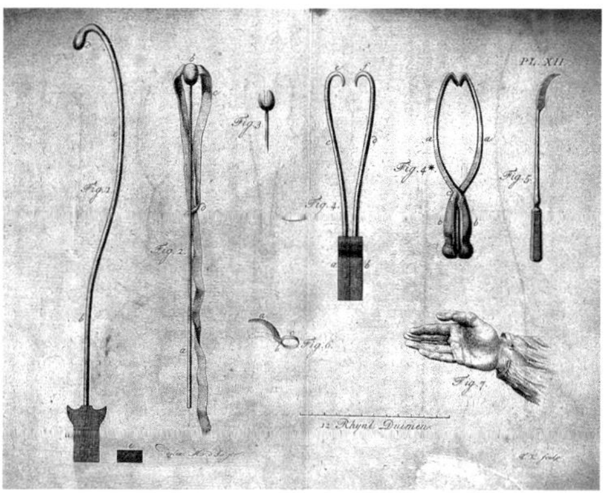

Johann Günther Eberhard, 1793, Wellcome Collection

Veterinary instruments from an 18th-century textbook. The one on the far right was used to file a horse's teeth and is identical to what is used today. The other instruments are more frightening and mysterious.

Cartorum

Postcard of the Royal Veterinary School in Lyon with a statue of Bourgelat in front. It appears to date from the late 19th or early 20th century and depicts the rebuilt and expanded facility. Little physical trace of Bourgelat's original school remains.

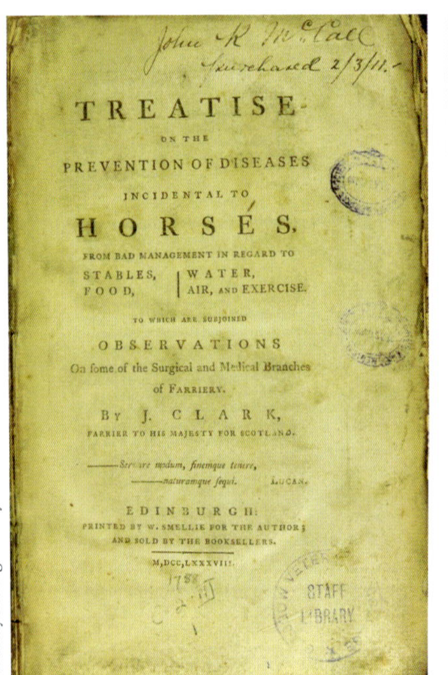

TREATISE

ON THE

PREVENTION OF DISEASES

INCIDENTAL TO

HORSES,

FROM BAD MANAGEMENT IN REGARD TO

STABLES, FOOD, | WATER, AIR, AND EXERCISE.

TO WHICH ARE SUBJOINED

OBSERVATIONS

On some of the Surgical and Medical Branches of Farriery.

BY J. CLARK,

FARRIER TO HIS MAJESTY FOR SCOTLAND.

——Servare modum, finemque tenere,
——naturamque sequi. LUCAN.

EDINBURGH:
PRINTED BY W. SMELLIE FOR THE AUTHOR;
AND SOLD BY THE BOOKSELLERS.

M,DCC,LXXXVIII.

Biodiversity Heritage Library

Cover of James Clark's 1788 treatise. The Latin quote below the title is from the Roman Stoic philosopher Lucan, and translates as "Observe moderation, be mindful of one's end, and follow nature." Despite his lack of formal education, Clark was clearly a learned man.

Dcs57, Wikimedia Commons

The blue plaque commemorating the Odiham Agricultural Society meeting in Odiham, Hampshire, England.

Royal Veterinary College

A 1795 engraving of Vial de Saint Bel teaching a farrier in front of the new London Veterinary College. That is allegedly "Ignorance" fleeing in the background on the left, chased away by the beatific and presumably learned Saint Bel. It is instructive to note the reflexive racism of the time, which portrays Ignorance as dark-skinned.

Web Gallery of Art

Eclipse, painted by Francis Sartorius in the 1770s. A handsome beast, but for some reason most artists right up until the 19th century usually made horses' necks too long and heads too narrow. According to various online polls, horses are still considered one of the top ten most difficult animals to draw, right up there with giraffes, elephants, and sloths.

E.B. & E.C. Kellogg, Harry T. Peters's "America on Stone" lithography collection, National Museum of American History

A lithograph from the 1840s of a fashionable woman and her equally fashionable pet canary.

BELOW: *A Couple of Foxhounds* by George Stubbs, 1792. Their eyes especially are painted in a way that makes it clear that the artist views the dogs as having emotion and personality.

Tate

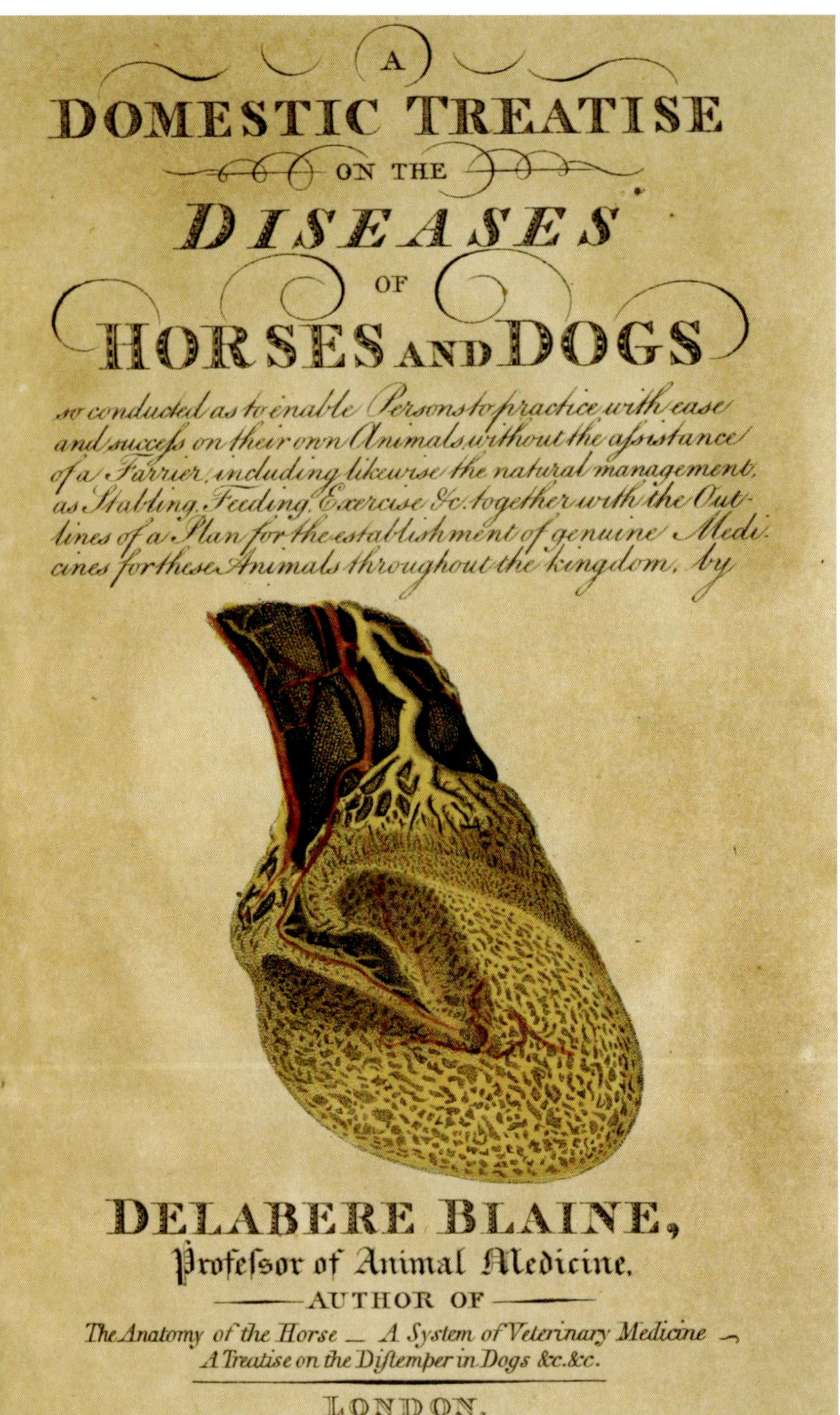

A

DOMESTIC TREATISE

ON THE

DISEASES

OF

HORSES AND DOGS

so conducted as to enable Persons to practice with ease and succefs on their own Animals without the afsistance of a Farrier, including likewise the natural management, as Stabling, Feeding, Exercise &c. together with the Outlines of a Plan for the establishment of genuine Medicines for these Animals throughout the kingdom, by

DELABERE BLAINE,

Profeſsor of Animal Medicine.

AUTHOR OF

The Anatomy of the Horse — A System of Veterinary Medicine — A Treatise on the Diſtemper in Dogs &c. &c.

LONDON,

Printed for T. Boosey, Old Broad Street, Royal Exchange.

1803.

The cover of the 1803 edition of Delabere Blaine's treatise. It is noteworthy that despite being focused on dogs in his own practice, the economic realities of the profession likely inspired his choice of cover illustration. He knew very well who buttered his bread.

Ontario Veterinary College

An 1891 painting by Paul Giovanni Wickson of a splendidly top-hatted Andrew Smith preparing to examine a horse.

College of Veterinarians of Ontario

The 1899 graduating class of the Ontario Veterinary College. I am so grateful that we were not asked to pose behind a bisected horse for our grad photo.

William H. Rau. Penn University Archives

The blacksmith shop at the Philadelphia Veterinary School in 1900. This is a callback to the days of the farrier vets a generation prior and would soon disappear from the curriculum. It's hard not to be impressed by the moustaches and ties, though.

A
DECLARATION
OF SVCH GREIVOVS
accidents as commonly follow
the biting of mad Dogges,
together with the cure
thereof,

BY
THOMAS SPACKMAN
Doctor *of* Physick.

LONDON
Printed for *Iohn Bill* 1613.

The Royal Society of Medicine Library

Title page of a treatise on rabies (London, 1613). As you have no doubt already surmised, Thomas Spackman, Doctor of Physick, sadly did not have the cure for the "grievous accidents as commonly follow the biting of mad Dogges."

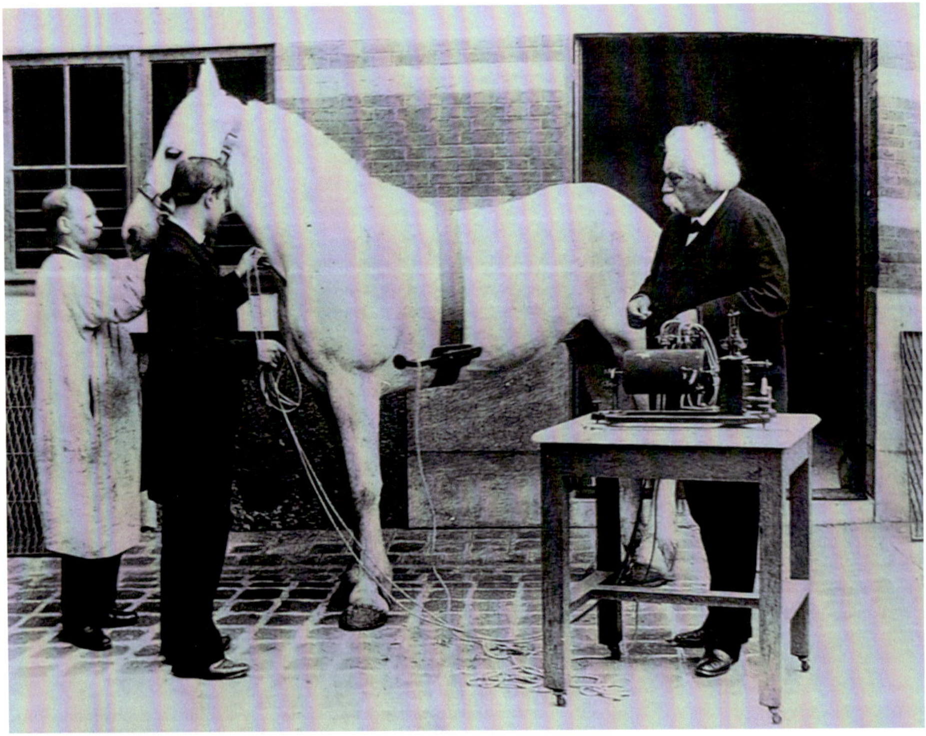

École nationale vétérinaire de Lyon

An early 20th-century demonstration by Jean-Baptiste Auguste Chauveau of his 1856 intracardiac cardiography experiment. Sensors have been placed in the right ventricle via the jugular vein, and the left ventricle via the carotid artery. One marvels at how calm the horse looks. Bored even.

Ulster Museum

John Boyd Dunlop (1840–1921).
A modest man with immodest whiskers.

The Museum of Art and Archeology Guéret

La visite du veterinaire by Sylvain Grateyrolle (1880), showing a veterinarian determining whether a sick cow has a fever by feeling her ear. This painting depicts the veterinarian in a more sympathetic light than Tschaggeny's *L'Empirique*. Grateyrolle also does a marvellous job of showing how worried the family is and how sick the cow feels.

Postgrad.ie

Aleen Isobel Cust (1868–1937). She does not look like someone to be trifled with.

Friedrichshain-Kreuzberg Museum

Maria Helene von Maltzan (1909–1997). The kittens' facial expressions are priceless — and immediately recognizable by any cat owner.

Horace Brown, Library and Archives Canada, e011067494-v8

Harry Colebourn and Winnie, Salisbury Plain, England (1914). Don't you love the way they're looking at each other?

Photo by author

The Tunnellers' Friends tympanum above the entrance into the Memorial Chamber in the Peace Tower. Parliament Buildings, Ottawa, Canada.

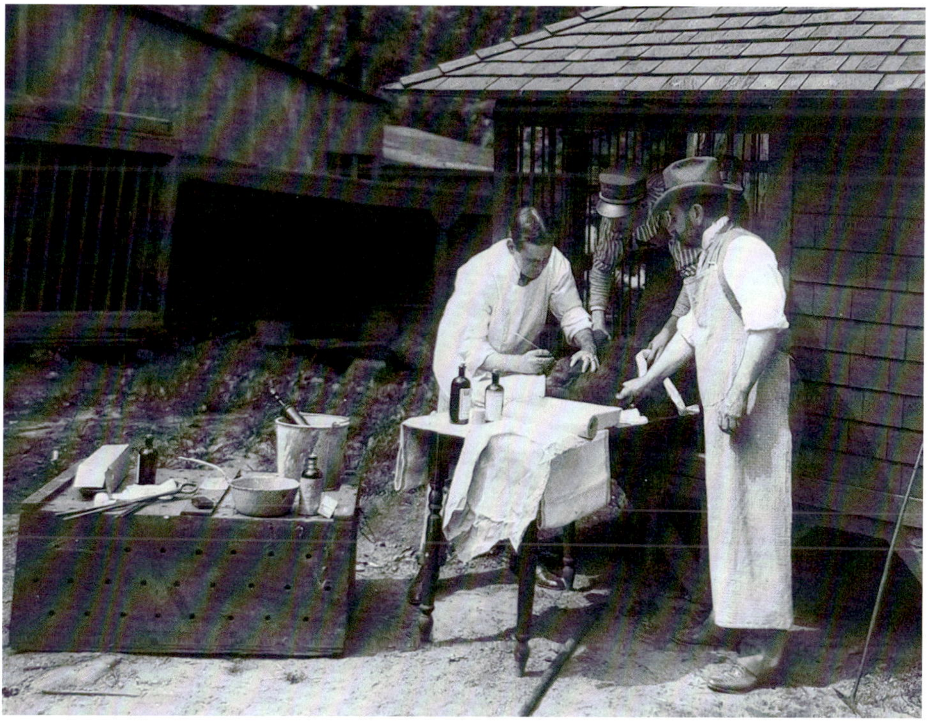

Wildlife Conservation Society

Dr. W. Reid Blair, the first zoo veterinarian in the USA, performing an operation on a Mexican grizzly bear (1904). I have questions. So many questions.

Stuart Thomson, City of Vancouver Archives, no. 506

The No. 6 Mobile Veterinary Section of the Canadian Siberian Expeditionary Force (1918). Note the handler on the bottom right holding a leash. His dog is seated with the officers. Another dog, half hidden on the far left, appears to have been of lesser rank, or just wouldn't sit as primly for a photograph.

James Alfred Wight (1916–1995), known and loved the world over as James Herriot, author of *All Creatures Great and Small*.

James Herriot Museum

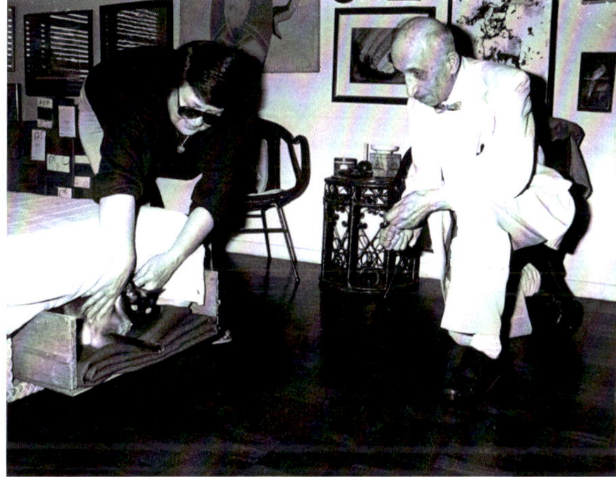

"Louis J. Camuti, DVM," Facebook

Dr. Louis J. Camuti (1893–1981). Gotta love the bow tie and white suit. I wish I could get away with that. And though that is a Siamese — a very popular breed in the mid-20th century — the other person is clearly not Olivia de Havilland.

New York Public Library

A veterinarian demonstrating the examination of a dairy cow at the New York World's Fair (1939). His shiny and clean shoes are a dead giveaway that this isn't happening on a real farm.

Sarah Marshall, courtesy of Conservation Through Public Health

Dr. Gladys Kalema-Zikusoka observing mountain gorillas in Uganda.

CHAPTER FOURTEEN

Moorcroft

William Moorcroft adjusted the tunic of his cavalry uniform and squinted at the sloping field. "If it is all the same to you, Major, Gibb and I are going to take a turn down by the pasture. Still getting our land legs, you know. It's glorious to be able to walk in a straight line for more than a dozen paces after all those months marching in tight tilting circles as we sailed down the Atlantic, round the Cape, across the Indian and then up the Hooghly."

Major Frazer gave a slight suit-yourself shrug and turned to speak to one of the servants who had accompanied the three Englishmen on their tour of the stables. Although he and Moorcroft were roughly the same age, Frazer looked much older, perhaps due to the rigours of life in a hot country far from home. Moorcroft guessed that he was pleased by this opportunity to return to the house to enjoy his late afternoon gin on the veranda and dream of Surrey.

Gibb, in a wool coat more suited to the North York Moors, mopped his brow and glanced furtively back towards the house.

Once Moorcroft and Gibb were far enough away not to be overheard, Moorcroft turned to his companion. "Did you ever imagine it would be this bad?" His tone betrayed a blend of bitterness and incredulity.

"No, I did not. Frazer may not be a veterinary surgeon, but even a fool knows that horses need to be kept with more space and light —"

"And better feed and water," Moorcroft said.

"And the horses themselves, quite aside from being disease-ridden, neglected, maltreated, and malnourished, are all wrong."

"Utterly. The mares are far too slight for cavalry purposes. And the stallions are hardly better. How could this have come to pass? What did Frazer think he was playing at? I'll write Leadenhall Street* in the morning. They will not be pleased."

"I can't imagine they will."

The two men arrived, sweating, at a fence at the end of the long sloping meadow that ran down from the stables. Beyond the fence lay a series of large paddocks. A few horses grazed there, indifferent to the strangers. England may be green and damp, but this was greener and damper than anywhere Moorcroft had ever been before. And the air had an unfamiliar heft and density to it, impregnated with a hundred unfamiliar aromas. It was not unpleasant, just unfamiliar.

They stopped and looked at the horses for a few minutes. Arab stock, Moorcroft thought.

Gibb spoke first. "If Boney** ever does show up with his heavy Norman steeds, this lot will be of no use."

"Even against the maharaja of Mysore's cavalry. The maharaja may have the same dainty horses, but I wager they are better cared for. Cornwallis's second war against him demonstrated that all too well." Moorcroft paused and kicked at the ground, as if testing the turf. "This is precisely why Leadenhall established this stud. Cavalry was sorely needed. And is now even more sorely needed. The invasion could come any time. But it is obviously not practical to ship all the animals from England."

* Headquarters of the East India Company; the company had recruited William Moorcroft to take over their Bengal stud farm, which was situated on a large estate outside of Calcutta.

** A popular mocking nickname for Napoleon Bonaparte.

"It was barely practical to ship us, sir," Gibb said with a wry chuckle.

Moorcroft allowed himself a broad grin. "Indeed. But now we're here and with this stud — we can do better. Much better. The company will have to be patient. It will take two years, at minimum. Maybe three."

"Where to begin?" Gibb had plucked a piece of grass and was about to chew on it but thought the better of it and let it fall while he waited for his boss to reply.

Moorcroft leaned on the fence, still looking straight ahead at the horses, while Gibb stood beside him. His words came out slowly, clearly drafting a list in his mind while he spoke. "First, we must attend to the medical needs of the horses. In the few minutes we were with them I saw sores, boils, rheumy eyes, lameness, and the whole devil's array of maladies. Enough to keep us busy a fortnight or more just addressing the most needful. Second, we must construct new stables, with far more air, light, and space. Third, we will have a better well dug, and we will plant some of these pastures with oats. We may need to have seed sent from England if nothing suitable can be found in India. And fourth . . ."

Gibb looked over at Moorcroft, waiting for him to finish, but it was if he had fallen into a trance, staring out ahead across the misty pasture to the three small mares grazing there.

"And fourth, William . . .?"

"Sorry. Yes, and fourth, we will need to obtain better breeding stock. Big solid animals. Heavier bone structure. I was just wondering where we would find them."

Two years later, in September 1810, William Moorcroft would finally figure out where. He wrote in a letter to his employers in London:

> The Cutch mares and several of those brought from the western and northwestern parts of Asia which have fallen under my notice are of such a size and frame as to prove abundantly that the countries which produced them are

> capable of producing horses very fit for the use of the army. On this account, it appears to me desirable to know how far the production of the Hon'ble Company's provinces nearest to the countries in question hold out a fair promise of being favourable to the breeding of horses.

This plan to find suitable horses in the "western and northwestern parts of Asia" would ultimately set his next great adventure in motion. The veterinarian was about to become an explorer, an aspect of his story which is well beyond the scope of this book, but which underlines that William Moorcroft is a remarkable figure in the history of veterinary medicine. It's astounding that he is not better known. There is only one, now out of print, biography of him,* and, tragically, nothing on film. Not yet anyway. In the meantime, permit me to lay out his life's story in a little more detail than I have done for the animal healers and veterinarians featured in the previous chapters.

William Moorcroft was born to an unwed mother on a farm in Lancashire, England, in 1767. Identified early on as a capable and highly intelligent child, he was taken under the wings of a trio of well-to-do gentlemen in the district who saw to it that he would achieve his obvious potential. This was at a time when inheritance of an estate, and the income that came with it, was the most common path to becoming a gentleman; that path was obviously not open to William. Instead, his benefactors financed his education as a surgeon, a profession that had relatively recently cleaned itself up after centuries of disrepute as barber-surgeons. Surgeons were now proper gentlemen, vying for respectability with their more customarily esteemed physician colleagues. So, off young William went to surgeons' school in Liverpool.

Let's call this his first adventure.

* Garry Alder's *Beyond Bokhara.*

Surgery at the time was, however, limited by the absence of anaesthesia and ignorance regarding the origin of post-operative infection. Ether wouldn't appear until 1846, and antisepsis, not until 1865 (thank you, Joseph Lister). Consequently, surgical procedures were either very quick or very nasty or, frequently, both. Nonetheless, Moorcroft applied himself with vigour. Later in life he would recall from those days the treatment of sailors coming home from Africa with Guinea worm. The technique was to slowly wind the parasite around a quill, extracting it from the patient's tissue (usually a blister on the lower half of the body) inch by inch. During his first two attempts he clumsily broke the worm, but "the serious consequences of this accident made me more cautious and more successful on subsequent attempts."

Then in 1788 (incidentally, the same year that James Clark published his treatise in Edinburgh), a pregnant heifer died on the estate managed by one of his benefactors, Thomas Eccleston. Pregnant heifers die all the time, of course, but this one would be the catalyst for Moorcroft's second adventure: becoming the world's first English-speaking, formally educated veterinarian. The heifer's death was quickly followed by two more. A cattle plague such as rinderpest was feared, and the whole district was deemed to be under threat. The local cow-leech was called in and immediately began bleeding the animals, smearing their nostrils with tar, jamming garlic into their crevices, and so forth. Eccleston considered himself to be a modern man of science, though, and looked on this with a skeptical eye. Then, in a flash of insight, it occurred to him to send for Moorcroft. Young William, then 21 years old, had been raised on a farm and was familiar with cattle, and was now trained as a surgeon, albeit for humans. Perhaps he could help.

And he did. When Moorcroft arrived, one poor cow had a terribly swollen throat and was in a desperate struggle to get her air, so William performed an emergency field tracheotomy. This brought the cow considerable relief, although it did not save her life. He also cooled fevered cattle with splashes of cold water and administered "Peruvian bark" (a source of quinine) in ale. Quinine was thought to be good for fevers in general; unfortunately for Moorcroft's patients, it turned out to only work on malarial fever (because it kills the parasite that causes malaria). Fortunately, although many others had apparently forgotten or neglected to rigorously apply Lancisi's and Bates's dictums from 75 years prior, Moorcroft knew to immediately isolate the herd and bury the dead cattle deeply. Consequently, the outbreak remained restricted to Eccleston's herd, and a larger disaster was averted.

This episode got Eccleston to add his voice to those beginning to push for the establishment of a veterinary school in England. More immediately, it prompted him to wonder whether in the meantime Moorcroft might be the perfect candidate to sponsor to study at the French school in Lyon. Moorcroft like the idea, stating, "If I were to devote myself to the improvement of a degraded profession, closely connected with the interests of agriculture, I might render myself much more useful to the country, than by continuing in one already cultivated by men of the most splendid talents." There were scores, probably hundreds, of successful surgeons in England, but not a single veterinarian. The choice for an ambitious young man willing to, as the modern cliché has it, "think outside the box" was obvious. Moorcroft would, in his words, "rescue a branch of medicine from a state of debasement."

Another of his benefactors, the celebrated surgeon Dr. John Lyon, was far less enthused. This is an understatement. Lyon thought that Moorcroft had tremendous potential as a surgeon and that switching to veterinary medicine was a substantial step down to a line of work widely considered low class and even repulsive. But Lyon was thinking of the veterinary medicine as had been practised by the cow-leeches, farriers, and folk healers. Eccleston and Moorcroft were looking ahead to a profession elevating itself with science and modern thought, no different in ultimate ambition than its human sister profession. However, this did not sway Lyon. He insisted that the matter be put to an independent arbitrator. Moorcroft's future was too important to be left to starry-eyed whim or passing fancy. The parties agreed to ask Dr. John Hunter, the then preeminent British surgeon, to consider both arguments and render judgment. After brief deliberation, Hunter was unambiguous in his opinion, stating, "If he were not advanced in years, he himself would on the following day begin to study the profession in question." Professional veterinary medicine's time had come, and most learned men could see that, even if John Lyon could not.

So William Moorcroft was sent to Lyon in March 1790. The Alfort school near Paris was better equipped by this point, but it was thought that Lyon would be safer, given what had transpired at the Bastille the previous summer. The period of study was somewhat flexible, originally up to four years when Bourgelat founded the school, but Moorcroft concluded all of his classes in a single hectic 12-month period, graduating March 7, 1791.

Along the way he seems to have become more interested in horses than cattle, and by 1793 we find him founding the first equine hospital in London, at 224 Oxford Street.* London at the time is estimated to have been home to 150,000 horses (and one million humans). The demand for horses was only going up and up, as was their value. Consequently, competent professional medical care was urgently needed. At the exact same time, however, a couple of miles to the east, in St. Pancras, the new Veterinary College of London was in crisis. Founded just two years prior, its first and only professor — the famous Vial de Saint Bel, whom we met in Eclipse's story — suddenly died. It's likely that he contracted the bacterial infection called glanders from the nasal discharge of a sick horse. Even today it can be challenging to treat if not aggressively attacked with antibiotics early in the course of the disease, and this was 150 years before antibiotics. Moorcroft was the most obvious candidate to replace the unfortunate Vial de Saint Bel, and negotiations went back and forth for several months, but he had invested too much in his new equine practice and, it seems, he feared he would chafe under the yoke of the college bureaucracy. This is another thing that hasn't changed. Ask any veterinary specialist debating between private practice and academic life.

Moorcroft's practice boomed. This was aided by the cachet of being the only qualified veterinarian and by the splendid physical set-up of the hospital, with a carriage archway into a large courtyard with ample stabling room — a rarity in central London. However, despite his formal

* Just east of Oxford Circus. A sporting goods shop is there now.

scientific training, the day-to-day reality of diagnosing and treating horses was not radically different than what the better farriers such as James Clark were doing. Not yet. Moorcroft still employed most of the delightfully arcane terminology for the diseases, such as mad staggers, phrenzy fever, rising of the lights, catarrh, thick wind, farcy, gullion, stag evil, pissing evil, pole evil (so many evils!), clap in the back sinews, and windgalls. And most of his treatments, including copious bleeding and a warlock's pharmacopoeia of dubious elixirs, salves, and drenches, would have been familiar to the leeches and farriers of centuries past. However, his education did put him in the frame of mind where he was willing to experiment and innovate. This was new. Pre-Enlightenment practitioners were slaves to tradition, reflexively honouring the old. For example, to address an otherwise intractable lameness, he surgically severed the nerve carrying the pain sensation. To everyone's delight, the horse was sound immediately after the surgery. Unfortunately, unbeknownst to Moorcroft, nerves can repair themselves, and the lameness returned after a few weeks. Incidentally, more than two hundred years later, resection of the nerve, albeit more thoroughly, remains a valid treatment option for a type of lameness called navicular disease.

As that anecdote illustrates, Moorcroft built his reputation on his skill and daring as a surgeon rather than on medical treatments. Sensible enough, as that was his original education. Recall that this was well before any sort of anaesthesia was available. Surgeons were almost entirely rated on their speed and accuracy. Success depended on getting things done as quickly as possible. It was the same, of course, on the human side. And Moorcroft was fast and highly skilled with the scalpel. He even operated on eyes to relieve pressure from glaucoma. Here is his description of the procedure (squeamish readers may wish to skip this passage):

> This operation requires some steadiness and address in the execution. The horse should be cast; and when the head is properly secured, the operator should, with the fingers of the left hand . . . press upon the eye from the eye-pit, so as to steady it, which however is not easily effected. With his

> thumb he should raise the upper eyelid, whilst an assistant presses down the lower; then holding a lancet with its edges standing upwards and downwards, and consequently the flat part towards the eye, he should push it into the clear part, at the outer corner, just before the white part, and carry it horizontally forwards; taking care not to give any other direction to the instrument.

This procedure would make most modern veterinarians nervous, even with the full benefit of 21st-century anaesthesia, instrumentation, and training.

Moorcroft also spent a considerable amount of time and much of his growing fortune* on trying to perfect the horseshoe. He wrote a book on the subject and, in 1800, was granted a patent for "A New, improved, and still more Expeditious Mode by the application of Machinery to my former Mode, of Making and Working Horse-Shoes," or, in short, a horseshoe machine. The machine, although it produced wonderful horseshoes, was itself too expensive to be a practical and economical solution, and there was widespread opposition among farriers to the automation of the ancient art of horseshoe making. So, it failed. Fortunately, Moorcroft's third adventure was just around the corner.

The same year as his horseshoe machine patent was granted, the East India Company approached Moorcroft. They were, as noted in the story at the beginning of this chapter, concerned both about the danger from attack by Indian princes and also about the possibility of a French invasion. Cavalry was key to success on the battlefield. The company had established a stud farm and wanted to engage Moorcroft to be their agent to purchase horses to add to the breeding stock. But the war with Napoleon was killing horses as fast as it was killing men, and horses were increasingly expensive. And shipping them, a dubious enough endeavour in peacetime, was fraught with dangers during war. So, in 1808, William Moorcroft agreed to travel to India to take over the operations of the stud

* Earning about two thousand pounds a year, or, to again employ our Jane Austen comparative scale, one-fifth of the income of the fabulously wealthy Mr. Darcy. Not bad at all for a young horse doctor.

and see what could be done to improve it without transporting horses halfway around the world. He had joined the Westminster Volunteer Cavalry in 1803 when the threat of Napoleon invading England was at its most acute, so he was able to address the Calcutta breeding operation with a cavalryman's eye. It should be noted that to lure him from his ever more lucrative private practice — the scarcity of horses drove up their value and what people were willing to pay for their care — the East India Company paid him three thousand pounds a year. Only the very top officials of the company were paid more.

After significantly improving conditions at the stud, Moorcroft's life becomes ever more that of an explorer and less that of a veterinarian. In fact, if you search his name, you will find far more references to his explorer side. The stimulus for this, his fourth adventure, was the search for sturdier breeding stock. This took him into Tibet in 1811, one of the very first Europeans to visit the closed country (the earlier ones had been unwelcome missionaries), and then in 1816 through Afghanistan to Uzbekistan, where he hoped to find the hardy Turkoman horses first described by Marco Polo. Among his many firsts, he and his companion were the first Westerners to see the giant Buddhas of Bamiyan.* Although it was at times against the East India Company's wishes (he even had his salary cut off for two years), he continued his explorations through Central Asia until his death by fever in Turkestan in 1825. The Napoleonic Wars had ended ten years prior with Napoleon's defeat at Waterloo, and Moorcroft never did improve the bloodlines at the Calcutta stud with the "Cutch mares" he dreamed of. But he is arguably a good candidate for "coolest veterinarian ever."

With that long digression into a single life over, let's turn our attention to the progress in small animal medicine in the early 19th century, and visit with a colleague of Moorcroft's.

---

* Destroyed by the Taliban in 2001 in one of the most egregious acts of archaeological vandalism in history.

CHAPTER FIFTEEN

Blaine

The knocking was at first so quiet that Delabere Blaine wondered whether it was the wind. He turned to go back to sleep. But then it became louder. It was unmistakable — someone was at the door. Edith was still asleep. She could sleep through almost anything. He climbed out of bed and lit a lantern. Finding his trousers, he pulled them on and threw a coat over his nightshirt before making his way down the stairs to the front door of the infirmary. He glanced at the clock on the mantel as he passed by: 1:15 in the morning. In the past, he would have been reluctant to open the door at this hour. Very reluctant. Georgian London at night had been an entirely different city, given over to vagabonds, prowlers, cutthroats, thieves, and disreputable persons of all description. New gaslighting had recently pushed this element to the remaining dark corners, but regardless, Delabere would have preferred to be allowed to sleep.

"Mr. Blaine," came an anxious voice — one he recognized but couldn't quite place — from the other side of the door. He opened the door to Geoffrey Youatt, his partner's brother, standing there holding a small whippet. William was out of the city, visiting his wife's family, otherwise Geoffrey would have surely gone to his home on Nassau Street instead. Delabere knew the man well enough to recognize him, but they were not otherwise well acquainted.

"Mr. Youatt, please come in. Who do we have here?"

"This is Sally. She has broken her leg. I am in the habit of walking her late. We were rounding a bend. She was run down by some madman. Racing by in a hackney." Geoffrey said this in bursts, panting and out of breath, sweat on his brow despite the cold spring night air. Delabere ushered him in and offered him a chair. Geoffrey sat. "The rogue must have heard Sally yelp and then heard me shout, but he didn't stop."

Delabere held the lantern over the slender, long-limbed grey dog cradled on Geoffrey's lap while her master spoke. He noted the odd shape of her left hip. It protruded higher and farther back than expected.

"That's dreadful, Mr. Youatt. This modern world makes one shake one's head, does it not? Please bring Sally to the table here so I can have a better look. I suspect this may not be as bad as it seems. Let us see. Perhaps we will be fortunate."

Geoffrey did as Delabere asked, and carried Sally into an adjacent room that had a large wooden table in the centre. He set her on the table. The dog was quiet and compliant, her large black eyes surveying everything around her but betraying no emotion beyond mild wariness. Delabere lit several more lanterns, washed his hands in a tin basin on the sideboard, then turned to gently palpate Sally's hindlegs, beginning with the sound one, before turning her over to address the injured one. Geoffrey winced, anticipating a painful reaction from his dog. But there was none.

"See, no grinding," Delabere said quietly.

Geoffrey's eyebrows went up.

"Her hip is simply dislocated, not broken. It moves smoothly and without significant pain, but it's not in the right place, so she cannot use it."

"Can you mend this?"

"With your help, yes."

"With my help?" This came out in a squeak.

Delabere ignored the man's doubts and explained, "The ball of the thigh bone that normally sits snugly in the socket of the hip bone has moved up and over. To here." He pointed to the protrusion visible under the skin. Like most of her kind, Sally was thin enough to make this obvious. "Do you see?"

Geoffrey nodded.

"So, now it is more geometry and physics than veterinary medicine. I must flex the leg so that the ball comes up." He made an upwards motion with his hand. "And then pull very stoutly on the leg at a sharp downward angle so that the ball pops back over the rim into the socket." He mimed the pulling action with both arms.

"And what of my help? What am I to do?" Geoffrey asked.

"You, sir, must hold your dear hound as tightly as you can. Holding against my pulling, so that all of my force goes into shifting the ball and not Sally."

Geoffrey nodded. He looked pale.

Delabere gave his client an appraising look and asked, "Perhaps a sherry first, Mr. Youatt?"

"Yes, please. That would be most welcome."

They drank their sherries quickly while Sally looked on, still perfectly quiet on the table. Once they were done, Delabere showed Geoffrey exactly how he wanted him to hold the dog. When he was confident that the still very nervous man knew what was expected of him, Delabere went around to Sally's head to pat her and speak a quiet word of encouragement. Then he went to her hind end and manipulated the dislocated hip. The injury was fortunately very fresh, so the muscles were still loose enough. He gave Geoffrey a curt nod and then, taking a deep breath, twisted the hip sharply and pulled with all his might on the leg, with the femur at as acute an angle as he could manage.

There was a soft pop. Sally didn't move or show any sign that anything had happened.

"Is it . . . in?" Geoffrey asked weakly.

Delabere moved the hip through its full range of motion. It was perfectly smooth, as if nothing had ever happened.

"It is." He looked at Geoffrey, who was now even paler. "You had better have a seat, Mr. Youatt. And another sherry."

Delabere Blaine's veterinary clinic on Wells Street* was just a six-minute walk northeast of William Moorcroft's place, although Moorcroft had already been in India for four years when Blaine started in 1812. Blaine was born in 1768 and got his introduction to veterinary medicine by acting as Vial de Saint Bel's translator at London Veterinary College. This position did not last very long, as Blaine clashed with the flamboyant French veterinarian: "Some impolitic attempts of mine to correct the anatomical errors of St. Bel made him wisely conclude that it would not be prudent to retain anyone about him who knew more than himself (which, as an anatomist, was little indeed) and I was, in consequence, dismissed." Blaine, as we will see, held himself in high regard.

He's probably best known today, inasmuch as he is known at all, as the author of the 1802 book *The Outlines of the Veterinary Art; or, the Principles of Medicine: As Applied to a Knowledge of the Structure, Functions, and Oeconomy of the Horse, the Ox, the Sheep, and the Dog, and to a More Scientific and Successful Manner of Treating their Various Diseases*.

It has the following marvellous dedication:

> To the King's Most Excellent Majesty.
>
> Sir,
>
> Your Majesty's gracious Permission so readily granted me to dedicate my last Work to you, emboldens me to lay this also at your Majesty's feet: and if that was thought in any measure deserving of Royal Approbation, I hope this will not be

* Number 5 Wells Street, at the corner of Eastcastle and Wells, a block north of Oxford Street. There's a pub there now, the Champion, which was established in 1860. The Wells veterinary infirmary appears to have closed in 1825 when Blaine retired and his partner, Youatt, moved all the work to his own Nassau Street practice.

esteemed less so, as having for its object the farther extension of a Subject as important as it is become popular. With the utmost deference and respect, I beg to subscribe myself,

Your Majesty's
Most devoted and most faithful Subject and Servant,
Delabere Pritchett Blaine.

The "last Work" he refers to is his 1799 *Anatomy of the Horse*. This new *Outlines of the Veterinary Art* is notable for being the first attempt to synthesize the history of veterinary medicine. It could therefore be viewed as the great-granddaddy of the book you are reading right now. It devoted a little over a hundred pages to veterinary history before going on to the anatomy and medicine of not only horses but, as the title states, cattle, sheep, and dogs. Subsequent editions of *The Outlines of the Veterinary Art*, through to the fourth and final in 1832, are interesting to compare to one another as Blaine steadily deemphasizes the roles of farriers and human physicians and surgeons, ultimately declaring veterinarians to be a distinctive "brotherhood." This seems obvious in retrospect, but it's important to remember how young the profession was at the time, and that most practising veterinarians were, like Moorcroft, from a human medical background and did not view themselves as being members of an entirely distinct profession but rather part of a more a junior branch of the pre-existing human one.

The veterinary profession was given its own charter in the UK in 1844, placing it under the governance of the Royal College of Veterinary Surgeons and thus putting it on equal legal footing with the analogous, but much older, Royal College of Physicians and Surgeons. Interestingly, though, Blaine himself never did become a qualified veterinarian in the sense of having completed a recognized educational program. His background was as a human surgeon's apprentice before coming to the Veterinary College of London, and it appears that although he took some courses there, he did not graduate. Keep in mind how new the college was, and consequently how dubious the qualifications it produced were in the eyes of many. However, Blaine clearly considered himself to be a veterinarian in every sense of the word, short of the quibbling technicality of his education. He was not a farrier, dog-leech, or moonlighting human surgeon. He was a veterinarian, full stop. And he has been recognized as such by the profession since then. In 1961, a laudatory note in the second volume of the *Journal of Small Animal Practice* recognized him as a pioneer of small animal medicine.

We can mark a significant advance with Blaine's next book, the 1817 *Canine Pathology, or a Full Description of the Diseases of Dogs; with Their Causes, Symptoms, and Mode of Cure . . . Interspersed with Numerous Remarks on the General Treatment of These Animals; and Preceded by an Introductory Chapter on the Moral Qualities of the Dog.*

Canine medicine had previously been given a few chapters here and there in the veterinary literature, including in Blaine's own previous book, but this book is one of the first, if not the first, dedicated entirely to the diagnosis and treatment of diseases in dogs. Well, almost entirely. He includes a three-page chapter at the end entitled "Diseases of Cats," which inauspiciously begins with the words, "Though these animals are inferior, in all their properties, to dogs . . ." The cat chapter is absent from his subsequent, otherwise expanded, 1824 edition.

Delabere Blaine was a dog person. In fact, in the preface he refers to himself as "the very father of canine medicine." Hmm. That argument can certainly be made, but it was still presumptuous, if not a little conceited, of him to go ahead and claim it for himself. Unlike Moorcroft's nearby exclusively equine clinic, Blaine's was focused on dogs, quite possibly making it the first "small animal clinic" in the world. Today approximately three-quarters of veterinarians work exclusively or predominantly with pets. Blaine and Youatt also treated horses, but far fewer than the recorded two to three thousand dogs per year. As previously noted, the demand for equine veterinary care was enormous at the time, and the value of the average horse was an order of magnitude higher than that of the average dog, so for economic reasons alone, an equine component to the practice remained important. Nonetheless, the 1812 establishment of the Wells Street veterinary infirmary for dogs is a landmark historical event, the bicentennial of which slid by unnoticed.

And who were these dog patients? Many were racing dogs, unfortunately, and dogs used for dog fighting and for the various baiting sports, especially bull and rat baiting. An excerpt from the October 1822 edition of *The Sporting Magazine* gives insight into the attitude and the times (I'm afraid that this is another passage that sensitive readers may wish to skip):

> Thursday night, Oct. 24, at a quarter before eight o'clock, the lovers of rat killing enjoyed a feast of delight in a prodigious raticide at the Cockpit, Westminster. The place was crowded. The famous dog Billy, of rat-killing notoriety, 26 lb. weight, was wagered, for 20 sovereigns, to kill 100 rats in 12 minutes. The rats were turned out loose at once in a 12-foot square, and the floor whitened, so that the rats might be visible to all. The set-to began, and Billy exerted himself to the utmost. At four minutes and three-quarters, as the hero's head was covered with gore, he was removed from the pit, and his chaps being washed, he lapped some water to cool his throat. Again he entered the arena, and in vain did the unfortunate victims labour to obtain security by climbing against the sides of the pit, or by crouching beneath the hero. By twos and threes, they were caught, and soon their mangled corpses proved the valour of the victor. Some of the flying enemy, more valiant than the rest, endeavoured by seizing this Quinhus Flestrum* of heroic dogs by the ears, to procure a respite, or to sell their life as dearly as possible; but . . .

And on it goes. At length. It's difficult for the modern reader to imagine that people got pleasure from such "sports," but it's not difficult to imagine how that might generate work for a veterinarian. However, many

* A misspelled reference to Quinbus Flestrin, Gulliver's nickname when he was among the Lilliputians.

of Blaine's patients were also simply pets.* As with Wilhelmi's canaries, these are animals looked after because they are loved and valued simply as companions.

Further on his fondness for dogs, you might be curious about the reference to "the moral qualities of the dog" in his book's title. Using numerous examples and anecdotes, Blaine goes on at great length about canine courage, fidelity, constancy, attachment, gratitude, and what he calls "sagacity." Blaine is quite taken by the intelligence of animals, generally, and also discusses the ability of bees to "compare, combine, and, perhaps, reason abstractly." It's an interesting and well-written chapter.

The medical chapters, however, are the expected mix of the useful, the useless, and the downright bizarre. The time of publication was still a generation before the revolution of germ theory, so his text is infused with the bunkum inherent with the old concepts of miasma and the humoral medical paradigm. Bleeding was still very much a thing, although Blaine tries to be sensible about it:

> The quantity of blood drawn should be regulated by the size of the dog: for a very small dog, one or two ounces are sufficient; for a middling sized dog, three or four ounces; and for a large dog, five, six, seven, or eight ounces, according to the size and strength of the patient, and the nature of the disease he labours under.

Various veins and approaches are described, concluding with this alarming statement:

> Or the tail may be cut in desperate cases; but, when this is done, it is better to cut off a small piece than to merely make an incision underneath; for I have seen, when this has been injudiciously done, the whole tail mortify and drop off.

* I'm a sucker for etymology, so if you'll indulge me, "pet" is first used in the 1530s in the sense of meaning an animal kept as a personal favourite. Originally the term was applied to lambs. It probably derives from "petty," as in "small."

However, the treatment of Sally's hip dislocation is taken directly from his writings and is exactly what we do today. Albeit with the benefit of anaesthesia and x-rays.

Blaine's principal claim to fame as a canine veterinarian was his approach to distemper. Even that 1961 article in the *Journal of Small Animal Practice* remarks that he will be "forever remembered by his masterly studies of canine distemper." That is a generous comment. Blaine does write about distemper at length, beginning by blaming continental Europe as the source of the disease as it was relatively recently recognized in Britain. He devotes a dozen pages to the description of the various forms of the disease, which was an advance, as even today it can be confusing to the inexperienced practitioner by presenting as either a respiratory, gastrointestinal, or neurological complaint, to name just the three most commonly afflicted body systems. However, his sometimes tiresome ego asserts itself again in his treatment recommendations:

> It is at this period of the disease I have experienced the happiest effects from the popular Distemper Remedy, discovered by me. This medicine has stood the test of nearly thirty years' trial; and although the varied appearances in the complaint render other auxiliaries absolutely necessary, yet no case of distemper can occur (that only excepted in which the purging continues without intermission) in which this Powder may not be given with great benefit in some stage of the disease.

Blaine's "Distemper Remedy" or "Powder" was a proprietary patent medicine, of top-secret composition, which sadly places him in the same dismal tradition as the infamous snake oil salesmen and other peddlers of dubious pharmaceutical concoctions. It was blatant hucksterism. It was nothing more than a money grab, given that he was otherwise very astute for the times, and had seen enough cases* to know that the "Remedy"

* Youatt kept careful statistics and reported 210 canine distemper cases in 1834, interestingly with almost all of them between March and October.

did not work. Nothing works. Not in 2023, and not in 1823. Not directly anyway. As we learned from Bonnie's prehistoric case, nursing care is key to boosting survival, plus possibly some medications (herbs in that case) to alleviate symptoms. This is much the same for most viral disease in humans today, such as COVID, influenza, and "stomach flu," to name just a few. However, to give Blaine credit, he then goes on to write:

> Likewise, when the Distemper Powders are not at hand, or when they have been tried without evident benefit, it will be prudent, after the directions already detailed have been complied with, to proceed with the following tonic plan of treatment alone; of which it is not too much to say, that it will prove nearly as universal in its application, and as salutary in its effect, as even the specific above alluded to:
>
> Gum myrrh .... 1 dram
> Gum benjamin .... 2 scruples*
> Balsam of Peru .... 2 drams
> Camomile flowers, powdered .... 1 dram
> Camphor .... 1 scruple.
>
> Mix with honey, conserve of roses, or other adhesive matter, into twelve, nine, or six balls, according to the size of the dog, and give one of them every night and morning. If the weakness becomes extreme, if the matter from the eyes and nose flows rapidly, and is very fetid, add two drams of cascarilla bark and a grain of opium to the mass of balls.

So, he admits that his proprietary Powder might be "tried without evident benefit," and he gives the reader a recipe for an alternative that they can mix together themselves after visiting the local apothecary.

* In the old apothecary system, a scruple is 20 grains, or a third of a dram. And a dram is an eighth of a fluid ounce.

Myrrh has some anti-inflammatory qualities, and benjamin, better known as benzoin, is a mild antiseptic. Balsam of Peru, a resin, might be antibacterial, and cascarilla bark might help with vomiting and diarrhea. But camphor definitely reduces coughing, camomile is good for the stomach, and opium is of course an excellent pain reliever, sedative, and anti-diarrhea medicine, among other properties. Long story short, Blaine's alternative recipe was quite reasonable, given what was available at the time, and given that the distemper virus itself could not (and cannot) be directly attacked.

Moreover, at the end of his discussion, he gives the best advice of all:

> The best preventive means that I know of, are to avoid or to remove all circumstances tending to produce debility, as looseness, low poor diet, too much exercise, exposure to cold, extreme evacuation from the nose, and, no less, the operation of mental irritation, from fear, surprise, or regret; all of which, I must again repeat, are very common causes of fits in distemper.

He is a bit off in believing that "mental irritation" is a very common cause of seizures in distemper, but he shows his humane side when he encourages dog owners to look after their dogs' general physical and mental well-being. Sometimes that's all you can do, and sometimes it's the best thing you can do. The reference to dogs feeling regret is sweet.

Before we leave Delabere Blaine, permit me one more passage from *Canine Pathology*:

> Spaying.
>
> THIS is a cruel and commonly an unnecessary operation, which is frequently practised to prevent inconvenience to the owners: but humanity should forbid its being resorted to, except in cases where the omission of it would endanger the life, as when some peculiarity occurs that would prevent

> a bitch pupping with ease and safety; or when she has been connected with, and is found to be breeding by, a dog much larger than herself. In this case, as she would probably die in labour, it is not improper to remove the puppies, at three or four weeks advance of pregnancy. The operation is performed by making an opening in the flank of one side, when the ovaria, being enlarged by pregnancy, are readily distinguishable, and may be drawn out and cut off, first one and then the other, securing the ends by a ligature lightly applied to each surface, leaving the threads without the wound. Farriers often apply no ligature, but content themselves with simply sewing up the wound, and no ill consequence seems to ensue. Bitches, after they have been spayed, become fat, bloated, and spiritless; and commonly prove short lived. Nature usually punishes any considerable deviations from her common laws; and it is observed, particularly among animals, that when the great work of propagation is artificially stopped, particularly in the female, she ceases to feel Nature's protection, and becomes diseased.

How times have changed. And how they are changing again. While the vast majority of female dogs are spayed in North America and the UK today, in Europe the percentage has always been lower, and in the last decade we have begun to postpone the procedure to adulthood to reduce the risk of certain orthopaedic issues. "Fat, bloated, and spiritless" is a dramatic overstatement, however, as any owner of wild and crazy spayed Labrador or Jack Russel will attest to. And the notion that the dog will become diseased once spayed because nature no longer views her as useful is, to be polite, far-fetched. Also of note is the aside that farriers spayed dogs in the past without ligatures. In other words, they removed the ovaries without tying off the ovarian arteries. Blaine claims that "no ill consequence seems to ensue." Astonishing. No veterinarian alive would dare something like that.* Perhaps dogs were made of sterner stuff centuries ago.

After his retirement in 1825, Blaine moved to the Isle of Wight where he lived the life of the prosperous country squire, evidently having earned enough from his Distemper Remedy, as well as the lucrative equine side of the practice. He died in 1845, one year after the foundation of the Royal College of Veterinary Surgeons. He must have been pleased to have lived long enough to see that happen. The epitaph at his grave reads, "Distinguished by the improvements he effected in the Veterinary Art."

* To be fair, in a large dog, ligating the ovarian artery can be much more technically demanding than the anodyne word "spay" implies. And although I stand by the statement that "no veterinarian alive . . . ," as recently as the 1970s, 150 years after Blaine's admonition, some older vets still removed ovaries without ligation. They counted on the trauma to induce muscle spasm in the artery walls and thereby prevent serious bleeding. Shockingly, this usually worked. Usually. And rumour has it that in some old-fashioned rural practices, tomcats are still today restrained for castration by placing them headfirst inside a rubber boot, exactly as Blaine described.

CHAPTER SIXTEEN

Smith

"The students are complaining," McEachern said. A faint smile played on his lips.

"Of course, they are, Duncan. That's what students do. One part studying. Two parts carousing. Three parts complaining. Isn't that what we did back in Edinburgh?" Smith said, smiling back.

"But they might have a point, Andrew. It's colder than a witch's teat in there."

Smith shrugged. "I'd offer to reschedule the dissection lectures to the summer. But then they'd complain about the stench of putrefying horses!" He paused and absentmindedly scratched his lavish muttonchop whiskers. "However, maybe tomorrow we can postpone the lecture to the afternoon. It should be a wee bit warmer by then."

"Thank you, Duncan. And hopefully the wind settles by then. It's shrieking up Yonge Street like a banshee right now."

"Good thing we Scots aren't afraid of Irish spirits," Smith said, chuckling. "Don't know about these Canadian and American boys, though."

He slapped McEachern on the shoulder. "Let's go in and give the complainers some instruction. The sooner they get through today's work, the faster they can repair to a pub and warm their delicate fingers."

McEachern nodded and smiled. He was seven years younger than Smith, and technically his employee, but he appreciated how he was treated as an equal. They were friends from back when they went to William Dick's veterinary school together, but in the hierarchical world of academia, these relationships were sometimes left at the office door. Andrew Smith might have seemed somewhat dour and old-fashioned to his students, but Duncan knew him to be warm and gentle at heart.

They stepped out of Smith's office and crossed a small courtyard to the wide double doors leading into what the two men half-jokingly referred to as their "Anatomy Theatre." It was, in fact, just an empty cavernous space Smith had leased for the purpose. It had the advantage of south-facing windows, so he could save on gaslight when the sun was bright.

"It *is* a touch nippy, isn't it," Smith said, hunching his shoulders against the wind. He noted the gaps around the doors and considered that perhaps the students had a point. McEachern swung them open, and they stepped inside.

Six long wooden trestle tables in two rows of three occupied the centre of the room. Around each table was a clutch of four to six students, all of them young men, all dressed in blood-stained smocks. Some were stamping their feet. Some had their hands tucked into their armpits. Each table had a pair of disarticulated horse legs laid on it, one hind and one fore. Smith mused, not for the first time, how ghoulish this scene would appear to an outsider wandering in off the street.

"Good morning, gentlemen," Smith boomed. He could see his breath. "I apologize for this uncommonly cold November morning. Much is under my control but, tragically, not the weather. However, we will make do. Work with speed, efficiency, and focus, and you will enjoy the dual benefit of warming yourself and being finished quicker."

There was a general nodding and shuffling of feet.

"You have all carefully studied your Stubbs*, I presume? Memorized the relevant diagrams? In which case, you will have no trouble identifying the superficial flexor tendon. This is your first task. Mr. McEachern and I will circulate among you and answer any questions you might have."

A hand shot up.

"Yes, Mr. Burrows? A question already?" Smith raised his eyebrows at a red-haired young man.

The student had a thick Bostonian accent and spoke quickly as if the words had been pressurized in his mouth in advance, just waiting for his lips to part. "And when I identify it, can I proceed directly to identifying the deep digital flexor tendon and the —"

Smith held up his hand to stop him. "Thank you, Mr. Burrows. Your energy and enthusiasm are always welcome. But no, please wait for Mr. McEachern, or myself, to verify your identification. The merest formality in your case, Mr. Burrows, I'm sure, but an important one. And please do permit your colleagues at the table to enjoy some of the pleasures of this dissection themselves. I wish we had more legs so that you didn't all have to share, but alas, the horses in Toronto are a healthy lot.** Perhaps in no small measure due to our own efforts here, gentlemen."

Burrows nodded. "Yes, sir."

"Now, if there are no further questions, prepare your scalpels, gentlemen, and begin to carefully incise the skin."

As the students busied themselves, Smith paced up and down the aisle between the tables and boomed, "The horse is the most important animal a veterinary surgeon will ever have the honour of treating. And the leg is the most important part of this most important animal. And the tendons in the lower leg are the most important part of this most important

---

* George Stubbs (1724–1806), an English artist most well known for his paintings of horses and dogs. He was also the author of *The Anatomy of the Horse* (1766), including a particular description of the bones, cartilages, muscles, fascias, ligaments, nerves, arteries, veins, and glands.

** Starting in 1890, students were required to buy their own horse carcasses. They could be obtained for three to five dollars, and up to ten students would share one. At that time, students also had to pay a five-dollar annual surcharge to use the dissecting room, on top of their 60-dollar tuition. This was the same tuition as at any of the three human medical schools in Ontario. To give a sense of scale, labourers earned about a dollar a day in that era.

part of this most important animal, so . . ." He stopped and held up his index finger to emphasize the coming point. "So, what you are doing this morning may well be the most important part of your education if you wish not to dishonour the initials VS, Veterinary Surgeon, that you will be privileged to put after your name, if" — he began pacing again and lowered his voice — "you somehow manage to pass this course of studies."

He spun on his heels and focused on a tall, thin, dark haired young man bent over his horse's leg in mid-dissection. "For example," he said, raising his voice, "Mr. Armstrong here will only manage to injure his colleagues and *not* manage to pass this course of studies if he insists on wielding his scalpel like a murder weapon!"

Armstrong straightened up and flushed red. He had been vigorously cutting towards the student who was holding the leg steady for him. The tissue was tough and frozen. Just as Smith walked by, the scalpel had jumped out towards the other student when it hit resistance.

"Mr. McEachern, could you please assist young Mr. Armstrong here by offering a brief summation of what has already been taught regarding proper handling of sharp instruments during autopsies and dissections?"

"With pleasure, Mr. Smith," McEachern said, and walked over to where the hapless Armstrong had stepped back from his dissection.

Just then there was a shout from the doors. It was Reginald, one of the stable boys.

"Mr. Smith, Mr. McEachern, come quickly, please! Mr. Bowes's horse has been injured. Grievously, it seems!"

Smith and McEachern exchanged a quick meaningful glance. Neither veterinarian had much use for the mayor, but his poor horse shouldn't suffer as a consequence of his owner's turpitude.

Andrew Smith had started the first veterinary school in North America — the first veterinary school in the western hemisphere, in fact. Upper Canada Veterinary School came into being on January 1, 1862, in a modest facility at 40 Temperance Street,* between Bay and Yonge in Toronto. Smith was born in 1834 in Scotland. He attended the Edinburgh Veterinary College, also called "Dick Vet" — now officially Royal (Dick) School of Veterinary Studies — after its illustrious founder, William Dick, where he met Duncan McEachern. Smith had planned to join the army as a cavalry veterinary surgeon, but upon graduation with high honours he was recommended by William Dick to recruiters from the distant colony of Upper Canada who were eager to find someone suitable to train veterinarians. The demand for qualified veterinary surgeons was growing rapidly in Canada, as it was everywhere.

Smith arrived in the fall of 1861 and within months welcomed his first students to the school, which then consisted of a brick house and a 300-square-foot stable. When he realized the scale of the task before him, he invited his old friend from Dick Vet, Duncan McEachern, to join him. The program started with two six-month sessions spread over two years. Attendance grew rapidly. Initially leased, Smith later bought the property in 1869 for $5,500. By 1875 he had run out of space and bought the adjacent building. In 1877, what was now called the Ontario Veterinary College announced:

> The new buildings now form the most convenient and commodious Veterinary Institution in America. The Lecture Room is large and well ventilated, and capable of seating eighty students. The Library, Pharmacy and Laboratory are complete and convenient. The Dissecting Room is 36 x 18 feet, and well lighted and ventilated. The Museum, when completed, will be the largest Veterinary Museum in Canada.

* For your reading pleasure, here is an irrelevant digression regarding the street name. Jesse Ketchum, a teetotaller who had made his fortune in the tannery business, donated the land the city wanted for a street to connect Bay and Yonge in 1837. His condition was that it be named Temperance Street and that liquor sales be banned along it. This held until late in the 20th century. As a member of the St. Andrew's Society, Smith is presumed to have enjoyed the occasional dram, but I'm sure he didn't mind having an additional excuse to forbid drinking at school. The site is now, predictably, occupied by a generic office tower. Smith lived across the street at 37 Temperance.

Two details jump out. The first is the implication that there were other veterinary institutions in "America"* by then. There were, and they were in competition with one another for students. More on that in a moment. The second is the reference to the college hosting the largest veterinary museum. There can hardly have been any competition for that title, so this seems an unnecessary boast.

The school was enlarged again in 1889, and then in 1922 it moved to Guelph, Ontario, about 90 kilometres to the northwest, in farm country where it made sense to be affiliated with the existing agricultural college. It's still there today, graduating 120 Doctors of Veterinary Medicine a year.

There may be a sharp-eyed Philadelphian or Bostonian out there whose eyebrows shot up at the declaration, several paragraphs ago, that Smith's school was the first in the western hemisphere. Indeed, the Veterinary College of Philadelphia preceded Upper Canada Veterinary School by a decade, receiving a charter in 1852. However, it failed to attract any students until 1859, when two applied, one of whom had already attended the Boston Veterinary Institute. To add insult to injury, before classes could get properly underway, all the professors resigned, apparently because they were expected to fund the school out of their own pockets. Eventually the Philadelphia Society for the Promotion of Agriculture stepped in, and some classes were held until 1866, when the hapless college closed for good. The number of graduates, if any, is unknown.

And what of the Boston Veterinary Institute? It was on more solid footing, but only operated from 1855 to 1860 and only produced a small number of graduates. The fact that one of them felt compelled to seek further education at the Veterinary College of Philadelphia is, perhaps, telling.

So, Lower Canada Veterinary School (now Ontario Veterinary College) is the oldest veterinary school in the western hemisphere *still in operation*. Or, you could say, the oldest *successful* veterinary school in the western hemisphere.

* Meaning all of North and South America. Until the Spanish-American War of 1898, "America" meant the whole New World, including Canada; only after did it come to refer solely to the USA.

What then is the oldest successful veterinary school in the USA? That honour goes to Iowa State College of Veterinary Medicine in Des Moines, which opened in 1879. The astute reader will note that this is almost a decade and a half after the end of the Civil War and might wonder how all the cavalry horses were looked after.* As alluded to in the story, several Americans went north to train at Upper Canada Veterinary School and, later, at a school Duncan McEachern opened in Montreal. Others went overseas. However, in total there were only 50 college-educated veterinarians in the entire United States during the Civil War, and most of them were in private practice. The Union Army saw the need for veterinary care but largely appointed untrained individuals to the role. The Confederate Army left the medical care of the horses and mules up to the individual soldiers and handlers.

Private veterinary schools kept sprouting up and failing around the US. In total there were around 40 of these, of predictably variable quality. There was fierce competition between them for students. As there were no national (let alone international) standards, some schools engaged in dubious practices. Some were little more than cash-for-degree factories, much like some of the shadier modern online universities. But others were honourable and provided a good education by late 19th–century standards. Even Smith's school was a for-profit business. He excelled in attracting students, often at the expense of his friend's school. McEachern's Montreal college struggled to fill its classes, largely because he insisted on a more rigorous scientifically based three-year program.** Prospective students often wondered why they should study harder and longer for the same "VS" qualification when they had other options. Upper Canada Veterinary School survived where the others failed not only because it was popular, but also because it hitched its star to the University of Toronto in 1897.

* According to the US National Park Service, about three million horses and mules served in the Civil War and, shockingly, approximately half of them lost their lives in the conflict.

** It operated from 1866 to 1902. Its most famous professor was William Osler, who was a human physician and would go on to revolutionize medical education at Johns Hopkins in Baltimore, Maryland. He is sometimes referred to as "the father of modern medicine." And he got his start at a little veterinary college in Montreal.

The future of veterinary education was through the university system, where it could be bolstered with public funds.

Elsewhere in the world, the first veterinary college anywhere outside of Europe opened in 1827 in Rashid, Egypt, in response to, what else, a rinderpest outbreak. The Rashid School of Veterinary Medicine quickly expanded its scope to horses as well, and ultimately offered a five-year program, far longer than the standard in Europe. Unfortunately, it closed in 1881 due to financial problems. Perhaps they should have consulted with Andrew Smith.

Coincidentally, that was the same year the first veterinary school opened in Asia. The forerunner of the modern Nippon Veterinary and Life Science University opened in Tokyo in 1881. It is described as having graduated "veterinary technicians"* rather than veterinarians, but the distinction, if any, was subtle in most of the world. Over in British India, Lahore Veterinary College was founded one year later.

And one more year on, in 1883, across the equator, the oldest veterinary school in the southern hemisphere opened: the Universidad Nacional de la Plata Facultad de Ciencias Veterinarias in Buenos Aires, Argentina. In 1888, Melbourne Veterinary College was the first in Australia. A successor program to the Rashid school started at the University of Cairo in 1901, so 140 years after Bourgelat, modern college level veterinary education was finally available on every inhabited continent.

* See NVLU's "Message from the President": https://www.nvlu.ac.jp/en/about-nvlu/001.html/.

CHAPTER SEVENTEEN

Galtier

"Hold her still, please, Jean," Galtier said, making no effort to disguise his irritation. Jean may have been one of the best students in the classroom, but he was useless at handling sheep.

Jean pursed his lips and tightened his grip, but the ewe continued to struggle. Galtier, who had been filling a large glass syringe with a clear viscous liquid, set the syringe down on the bench and stepped up to the young man and his reluctant research subject.

"Like so," Galtier said as he bent forward and used his right arm to push the sheep firmly against his right leg. Then he pulled up her chin with his left hand so that the back of her head was against his chest. "You must demonstrate confidence," he said as he straightened up again. "It is not force of body that persuades the animal, but force of mind."

Jean nodded, although his eyes betrayed a flicker of confusion, and tried again. The sheep was not as still as she had been with the professor, but she was more settled than before.

"That will have to do," Galtier said. He picked up the syringe again and checked the level against the light from the high windows along the far wall of the laboratory.

"Okay, girl," he said softly as he slid the needle under the loose skin of her chin. "This is the last time." He patted the sheep on her head when he was done.

"That was four times, yes?" Jean asked.

"Seven. But you have only seen four. Alexandre has been helping as well." Galtier walked over to a large chipped enamel sink and began to wash his hands. "You may return her to the barn now. Thank you."

Jean picked up the rope lead and coiled the loose end in his hand slowly. He cleared his throat and glanced over at Galtier, who was now done washing his hands and was unselfconsciously smoothing the ends of his waxed moustache. "Professor, you said this was the last time. What is the next step? A challenge?"

Galtier stopped fiddling with his moustache and turned to face his student. "You are correct. Next week we will challenge our ovine friend with the saliva of a mad dog."

"And you do not expect the sheep to go mad?"

Galtier smiled. Despite his irritation at Jean's poor livestock handling skills, he liked the young man. He had a more curious mind than most of the students at the Lyon school, who were there either because their fathers wanted them to be or because they simply wanted a good job and had rejected the other professions or been rejected by them. "Mad?" he answered. "What do you know of rabies in sheep, Jean?"

"Sorry, sir, I misspoke. Sheep acquire the same dumb form of the sickness as cattle. They drool excessively and become somnolent and eventually paralyzed."

Galtier continued to smile and nodded, encouraging Jean to go on. Noting this approval, Jean allowed himself a little joke. "They do not run about and bite people or other sheep."

"No, they do not! But to answer your question, I do not expect the sheep to become infected with rabies. And why is that? You have not been taught this as it is very new, but given what you've seen here, what do you think?"

Jean bit his lip and looked at the sheep and then at the vial Galtier had pulled the clear liquid from.

"That is rabbit saliva, is it not? I know you have infected rabbits with rabies as well."

"That is correct. It is saliva from infected rabbits." Galtier looked like he was about to say something more but stopped himself. Let the student follow the thread.

"But after six times, why is this sheep still healthy?"

"Why indeed." Galtier grinned and began playing with the tips of his moustache again.

"I can only suppose that the rabbit saliva is not as strong as dog saliva. I have never heard of a person contracting rabies from a rabbit. Although most people would not allow themselves to be bitten more than once by a rabbit, let alone six times!"

"That is also correct. You are following the correct scent, Jean. For reasons as yet mysterious to modern science, the infective rabies particle becomes weaker or attenuated in some animals, such as rabbits."

"And you believe," Jean began slowly, "that exposing the sheep to these weakened virus particles will protect her against the true disease-causing ones from the dog?"

"Precisely. Yes, that is what I believe."

"Like Monsieur Edward Jenner in England 80 years ago.* The rabbit rabies is as the cowpox was." Jean spoke faster and more confidently now.

"Very much like that. Very good, Jean."

"A vaccination against rabies!"

---

* In case you're unfamiliar with the story, in 1796, after observing that smallpox never seemed to affect milkmaids, Edward Jenner injected an eight-year-old boy with cowpox and then two months later injected him with smallpox. Cowpox is very similar to smallpox but does not make people ill. The boy survived. A friend of Jenner's coined the word "vaccine" for his procedure, deriving it from "vacca," Latin for "cow." For centuries prior, people all over the world, from Africa to China, had been using very small doses of smallpox to try to provide protection against future outbreaks. While this did work, it had obvious risks. Finding a safe way to fool the immune system was the breakthrough.

"I certainly hope so. In a few weeks we will know. Rabies from a dog bite is uniformly fatal in sheep, so I expect the sheep hopes for success as well."

The heroic sheep lived, although what became of it after the experiment is unrecorded. Pierre Victor Galtier, veterinarian and professor at the veterinary school in Lyon, had taken a major step in defeating one of the greatest zoonotic scourges in history. Because it killed its victims so quickly, and because humans were "end stage hosts" (i.e., can't pass it on to others), rabies never caused epidemics or killed very large numbers of people.* However, it is a horrifying disease that has been prominent in human imagination and culture for millennia. It was horrifying because every person afflicted by rabies died. Every last one. Some people survived the plague. Some survived consumption (tuberculosis). Some survived smallpox, diphtheria, influenza, and the whole rest of the devil's pantheon of infectious diseases. But nobody survived rabies. And on their way to their inevitable end, they suffered horrendously. It was terrible to behold the torment of someone driven to extreme agitation and hallucinations by rabies attacking their brain.

Consequently, a lot of thought and effort were put into preventing and treating the disease. You'll recall the number of ancient and medieval proto-veterinarians who proposed rabies cures, ranging from cutting the "worm" under the tongue to putting the anus of a rooster over the wound. All of it either useless or worse than useless, with the exception of cauterizing wounds immediately after an infective bite. This might have prevented a few cases if done aggressively enough. But a vaccine had the potential to prevent every case. Not only for rabies, but also for most viral diseases. After the success of rabies vaccination, one viral disease after another succumbed. The reason there was such a long gap between Jenner's work with smallpox and the breakthrough with rabies

* Currently about sixty thousand people a year die of rabies worldwide, which is still more than a few too many. But compare that to the one to two million deaths from dysentery.

was that there was no cowpox version of rabies, or for most of the other serious viral diseases. The relationship between cowpox and smallpox is highly fortuitous. Both prompt an immune response in humans that protects against infection by either in the future, but only one can make you fatally sick. The innovation with rabies was the concept of attenuation — passing the virus through another species to weaken it. Many vaccines today are still produced this way, for example, influenza, which is attenuated by injecting it in a series of chicken eggs. In 1879 Galtier noticed that rabies became weaker after infecting rabbits. This was the basis of his 1881 sheep experiment.

Now, some of you will have noticed that a name is missing from this discussion. A famous name inextricably linked to the history of vaccines generally and of rabies vaccines specifically. That name is Louis Pasteur. Most casual references to the discovery of rabies vaccines, as well as some quite serious and detailed references, talk only of Pasteur. Galtier is not mentioned. This is in part because it was Louis Pasteur who performed the first rabies vaccination on a human in 1885. And it is in part because Pasteur was a shameless self-promoter who made use of Galtier's pioneering work and quite consciously failed to credit him publicly for it. Galtier wasn't Pasteur's only victim. Several other 19th-century scientists contributed to Pasteur's successes and were ignored.* This is not to say that Louis Pasteur was not a great man. He clearly did marvellous things for humanity. We should all be grateful. It's only to say that the drastic imbalance between his fame and Galtier's is, to say the least, unfair. Pasteur's name is everywhere, whereas Galtier has one street named after him in Lyon, a bust at the veterinary school there, a handful of scholarly articles about his work, a short Wikipedia article, and now this chapter in my book.

During his life, however, he was well regarded by his colleagues, and many looked askance at Pasteur's grandstanding. Galtier was even considered for the Nobel Prize in Physiology or Medicine in 1908, but, unfortunately, he died that spring, and it cannot be awarded posthumously.

* If you're a Pasteurophile and are distressed by this news, I refer you to Jean-Marc Cavaillon and Sandra Legout, "Louis Pasteur: Between Myth and Reality," *Biomolecules* 12, no. 4 (2022): 596, https://doi.org/10.3390/biom12040596.

Pierre Victor Galtier not only merits a chapter because he is an unsung veterinary pioneer, but also because his story highlights the fact that treating sick animals is not the only career path available to veterinarians. Many, then as now, work in research and teaching as well as regulatory bodies and industry. In the US in 2020, 83 percent of vets were in private practice, 8 percent in universities, 5 percent in government agencies, and 4 percent in industry (largely pharmaceutical and animal nutrition)*. Galtier had been the top student in his class at Lyon, so perhaps academia appealed to him, but he started his career in private practice, working for a veterinarian in Arles, France. As an interesting aside, he married the boss's daughter before leaving the practice to begin teaching pathology and microbiology at his alma mater. As more veterinary schools were opened, more veterinarians were needed to serve as faculty, although many human physicians, and people from other allied disciplines, taught in veterinary schools as well. To a sharp young veterinarian with a curious mind, the steady paycheque of a university job and access to its research facilities held a lot of appeal. Some schools, such as Smith's Upper Canada Veterinary School, did very little research, but others, such as the National Veterinary School in Lyon, afforded their faculty ample opportunity to pursue areas of scientific interest.

Galtier was among the first in an honour roll of veterinarians who made their mark in scientific research, in many cases with crossover benefits for human medicine. Other prominent veterinary scientists helping humans include Bernhard Bang, the Danish veterinarian who in 1897 discovered the cause of brucellosis, also called undulant fever in humans, and Vanessa Hirsch, a renowned AIDS researcher still active today.

A curious member of this elite group of veterinary innovators was John Boyd Dunlop, an Edinburgh graduate who ran the largest veterinary practice in Ireland in the 1870s and '80s. One day in 1887 he was watching his son bumping along uncomfortably on his tricycle when he had a brain flash. Using a sheet of India rubber that he kept in the clinic

* American Veterinary Medical Association, "U.S. veterinarian numbers 2020," https://avma.org/resources-tools/reports-statistics/market-research-statistics-us-veterinarians-20.

for various purposes,[*] he fashioned a tube with a valve to wrap around the wheel. Then he filled it with air. Dunlop had invented the pneumatic tire. Within a few years, sales of Dunlop's tires exploded with the boom in cycling. In 1895 he sold his interest in the tire business at well below market value and remained in quiet retirement dabbling in, of all things, drapery, until his death in 1921. Dunlop was a modest man who preferred to avoid attention. Having campaigned successfully for better milk and meat hygiene legislation, he always considered his contributions to veterinary medicine to be more important than having spawned a massive new multi-million dollar industry.

Scientific research wasn't the only domain where veterinary medicine made significant advances in the late 19th century. As was occurring elsewhere in society, signs of progress were also finally showing in the breaking down of gender barriers in the profession. In 1889, while Galtier was working on rabies and Dunlop on tires, a Miss Stephania Kruszewska graduated from the veterinary school in Zurich, Switzerland. Unfortunately, nothing is known about her other than that she was Polish, presumably having studied in Switzerland because there were no veterinary schools in Poland. We also know that she returned to Poland upon graduation, but beyond that, the record is silent.

After Kruszewska, there are scattered references to other female graduates from a variety of schools, but in most schools, most years, the graduating classes were entirely male.[**] But let's visit with the eighth female graduate we know of, Elinor McGrath, who opened her small animal clinic in Chicago in 1910, the first woman to practise veterinary medicine in America.[***]

* In the 19th century, veterinarians often had to make many of their own medical devices. Rubber was especially useful in all kinds of straps, cuffs, catheters, and tubes.

** Almost 50 years later, in 1938, there were still only 21 female veterinarians in the whole US.

*** In addition to Kruszewska, the other six before McGrath were two more Polish students and one Ukrainian in Zurich around 1889; Marija Kapčevič, also from Ukraine, who graduated from Alfort, France, in 1896; Aleen Cust, from Edinburgh, in 1900; Mignon Nicholson, from Chicago, in 1903, who did not go into practice; and Belle Reid, from Melbourne, Australia, in 1906. More on Cust and Reid anon.

CHAPTER EIGHTEEN

McGrath

"I'd like to speak to the veterinarian," the man said as he removed his cap and shook off the rain. He had come in with a German shepherd, who sat beside him in the water pooling on the floor. McGrath was impressed that he didn't shake himself dry all over her tiny waiting room. This is a well-trained dog, she thought.

"Dr. Elinor McGrath," she said. She smiled and stepped forward to shake the man's hand. "Pleasure to meet you, sir, and your handsome dog."

The man frowned. "That's very funny, young lady, but I doubt your boss would approve of jokes as a way of greeting new clients. I'm a busy man, so please don't waste my time. Just fetch the doctor for me."

"I am not in the habit of sharing jokes with people I don't know well. I am the doctor. Fully qualified. Now, what may I do for you and your dog?" McGrath continued to smile as she said this, but it was now a rigid smile. This sort of thing happened every day. Every day. She was still trying to learn the best way to approach the skepticism and outright disbelief.

The man stared. His mouth opened slightly, and then he closed it again. "That's not possible," he finally said. He shifted slightly and glanced around as if trying to find evidence to refute this bizarre assertion by the young woman standing in front of him.

McGrath kept smiling. "Did you want to see my diploma? Or the testimonials clients have written for me?" She reached for a leather journal on the reception counter.

The man held his hand up. "No, that won't be necessary. My neighbour said that there was a new vet on Indiana Avenue, but he didn't mention . . . Anyway, thank you for your time. I'll go back to old Doc Cooper. He's more of a horse man, but he does well enough with dogs."

"I understand, sir. I'm sure you'll feel more comfortable having this handsome fellow in familiar hands. What's his name?" She crouched down and extended her hand towards the dog.

The dog, who had been sitting rigidly by the man's side, staring straight ahead, turned his head to look at McGrath's hand.

"His name is Prince. Be careful, young lady. He doesn't take kindly to strangers. Even took a piece out of Doc Cooper once, and he knows him."

McGrath nodded and then spoke softly to the dog. "It's okay, Prince. I'm just saying hello. Maybe I can offer you a little piece of beef?"

She straightened up. "Would that be okay, Mr. . . .?"

"Mr. Doyle. Joe Doyle. Just a small piece, mind. And watch your fingers."

"I picked it up fresh from the butcher this morning on my way in." McGrath stepped behind the counter and pulled out a small wax paper packet tied shut with string. She deftly untied the string and began sorting through the contents of the packet.

"You're the first dog this morning, Prince, so you can have the choicest morsel. How does this piece look?" She held up a thumb-sized piece of beef. Prince looked up at it and began to pant, but he remained otherwise motionless.

McGrath stepped back around the counter and crouched again. "Here you go boy."

Prince glanced up at Doyle, who nodded. The dog very gingerly took the meat from McGrath's flat palm. He swallowed it in a single gulp without chewing and started wagging his tail. Then he leaned forward and licked McGrath's palm.

McGrath laughed. "What a good boy you are."

"Well, I suppose he can be partial to the ladies." Doyle paused as if gathering his thoughts. "But we'd best be off now."

"Thank you for stopping by. Good luck."

"Goodbye, and best of luck to you as well."

Doyle put his hat back on, turned up the collar of his coat, and walked towards the door, Prince at his heel.

Prince stopped and sat down again. Before Doyle could say or do anything, the dog stretched his neck out and made a violent gagging sound, as if he was trying to clear something from his throat. But nothing came out.

Doyle turned. "This is what he's been doing. Several times a day. It's getting worse."

"But we saw him swallow the beef," McGrath said, brows knit with worry.

"He did. It always sounds like he's choking, but I don't think he is because it happens at random times. Mostly not when he's eaten."

"Let me have a look, Mr. Doyle. Please."

The man stood at the door, looking uncertain. Then Prince gagged again, even more dramatically.

"Okay, just a look. Especially since he might not retch for Cooper," Doyle said.

"Thank you. Now, what I need you to do is crouch beside him and brace him against you with your elbow while you open his mouth for me. He seems to be a very well-behaved dog."

"He is. He's the best dog I've ever had. I don't need to brace him or anything like that. If I tell him 'open,' he'll let me do it without flinching."

McGrath nodded. "Good. I'm going to fetch a lantern."

"You need to buy one of those new battery-powered lights."

"When I have a few more patients, I might be able to buy a few more things for the practice," she said as she lit a lantern that had been sitting on a side table. "Now open his mouth, please."

Doyle said, "Open," and Prince relaxed his mouth and allowed his owner to pull his jaws wide apart.

McGrath bent forward, lantern in one hand, and moved Prince's tongue aside with the other. "I'm so glad he's a good dog," she muttered to herself.

She peered into the mouth for what seemed like a very long time. At one point she reached into the throat with a finger. Prince gagged slightly, but otherwise remained still.

"Okay, you can let him go," she finally said. "Good boy, Prince."

"Did you see anything?" Doyle asked.

"It's his tonsils, Mr. Doyle. They look like a pair of ripe cherries, red and swollen."

"Tonsils? Dogs have tonsils?"

"Dogs have everything." McGrath chuckled as she wiped her hands on a towel.

"What can be done? Do you know?"

"They'll have to come out. Just like in a person."

"Come out? You can't be serious."

"I am. I have performed the procedure successfully myself."

"This is too much." Doyle shook his head. "I thank you for your time and opinion, but I assume you won't take offence if I seek a second opinion from Dr. Cooper."

"No offence would be taken at all."

Doyle nodded curtly. "What do I owe you for this?"

"Nothing, Mr. Doyle. Consider it a courtesy."

"Thank you. Good day." And the man and dog stepped out into the rain.

The veterinarian wiped her hands again, returned to where she had been sitting behind the counter, and waited to see who would come in next.

Far too little is known about Dr. Elinor McGrath. We know that she was born in 1888, although not what date, and we know that after graduation in 1910 she practised at her Indiana Avenue, South Side Chicago, small animal clinic for more than 30 years. After that she worked as the Illinois State Veterinarian. She was also the first female member of the American

Veterinary Medical Association. A few other snippets have trickled down. One is that she was known for decorating a clinic Christmas tree with pet treats every year, and another is that she founded the first pet cemetery in Chicago. And yes, she did perform tonsillectomies in dogs. According to Ivan Katić's excellent paper, she was the first veterinarian in the world to do so.*

That we don't have more information about the first women in this now female-dominated profession is terribly sad. But also, unsurprising. Katić's paper is the only readily available reference that provides any level of detail. And as of 2024, it had only been cited four times. None of the Eastern European female veterinary pioneers is mentioned anywhere else in more than a brief sentence. The only early female veterinarians we know much more about from multiple sources are Belle Reid and Aleen Cust, both of whom even have Wikipedia pages. And Cust had a short biography published, which is unfortunately now out of print.

Both Reid and Cust appear to have been interesting people, so they deserve some space here. And although she was born a generation later, we'll also spend some time with the Countess Maria Helene von Maltzan, veterinarian — and resistance fighter against the Nazis.

But first to Australia, where Isabelle (Belle) Bruce Reid was born into a well-to-do family in Melbourne in 1883. Her first passion was singing, which she excelled at, but her parents did not consider this to be an appropriate career path for a young woman of her social standing. Remarkably, however, they had no qualms about supporting her plan B, veterinary medicine. In 1902 she enrolled in Melbourne Veterinary College, the first woman to do so. Four years later, she was the only student to pass the final exams. On November 21, 1906, she became the first woman in the world to officially register as a veterinarian. She set up practice on Whitehorse Road in Balwyn, then a small town, now an eastern suburb of Melbourne. Balwyn Veterinary Surgery is still in operation today in the same location with, suitably enough, an all-female veterinary staff. Mind you, the majority of Australian veterinarians are now women, so it's not much of a surprise.

* Katić, "Pioneer Female Veterinarians," 145.

Sadly, Reid retired early in 1923, in large measure because her male colleagues refused to accept her. After leaving practice she turned her focus to breeding Irish cob horses, Jersey cattle, Pomeranians, and Irish wolfhounds. The combination makes for an arresting mental image. She also played polo.

Halfway around the globe, Aleen Cust had been in practice for several years before Reid, but without official recognition. She was born in Ireland to minor nobility* in 1868. Unlike Belle Reid's parents, Cust's did not approve of her plan to study veterinary medicine. Remember that veterinary medicine at the time was still very much associated with farriery in the public imagination, and consequently had a "lower class" taint. But Cust did it anyway. Her one concession to her family's opprobrium was to change her name to Custance when she applied to the New Veterinary College** in Edinburgh in 1895. A mere two years later, when it came time to sit her first professional exam, the Royal College of Veterinary Surgeons (RCVS) refused to allow it. Cust appealed this decision, but the Scottish Court of Session refused to intervene, stating that the RCVS was not "domiciled" in Scotland, so the court had no jurisdiction over it. Not domiciled, yet still having authority over veterinarians there. Cust chose not to pursue the obviously illogical paradox to a higher court in London but instead completed her studies in 1900 without sitting the licensing examinations, and then she went back to Ireland. There she had the good fortune to find a veterinarian, William Byrne, who was happy to employ her as an assistant. Interestingly, the dean of New Veterinary College, Professor William Williams, wrote her a letter of recommendation. He was apparently no fan of the RVCS's fusty rules. Nor was William Byrne, who allowed Cust to fully practise as a veterinarian, lacking only the title. Ireland was still part of the UK then, and the RCVS had theoretical authority, but Byrne and Cust didn't care. Nor did, apparently, many of their clients. When Byrne died in 1910, Cust took over the practice.

---

* Her father was 2nd Baronet, Sir Leopold Cust. After Sir Leopold died, Cust's mother, Lady Isabel, became Woman of the Bedchamber to Queen Victoria. Later, her brother was appointed equerry to King George V.

** Rival to the venerable Royal Dick Veterinary College, only 100 metres down the road. New Veterinary College moved to Liverpool in 1904, eventually becoming the University of Liverpool Faculty of Veterinary Science.

In 1914 she volunteered to serve on the front in the First World War, driving her own car to Abbeville, France, where she treated injured horses for the Army Veterinary Corps and Remount Service. This service, plus the 1919 passing of the *Sex Disqualification (Removal) Act*, paved the way for Aleen Cust to finally be allowed to apply to become a member of the RCVS. Perhaps as a sign of contrition, the RCVS recognized her years of experience and only required her to pass the oral part of the examinations. She passed, of course, and on December 21, 1922, became the first female member of the world's most august veterinary organization. One interesting vignette regarding Aleen Cust — she apparently made many of her farm visits riding side-saddle on a white Arab stallion.

Maria Helene von Maltzan, the last of our trio, was also from a wealthy background. This is an important point. The professional advances made by women in the late 19th and early 20th centuries were largely only available to the upper classes, whose daughters had the leisure to consider a much wider range of career options. And, more importantly, who had the luxury of the family safety net, making it so much safer to defy societal norms. The Countess von Maltzan was the youngest of eight children. She was indulged by her father and permitted to range freely around their enormous estate in Silesia (then Germany, now Poland). There she developed a passion for animals. According to Ulla Schweer's biography of von Maltzan, she almost drowned her brother for killing snakes, and she stabbed another boy with scissors after he cut off a newt's legs.* Passion, indeed. Unfortunately, her father died when she was 12, and she didn't get along with her mother, so she was sent away from her rural paradise to a boarding school. There she plotted an educational path that would eventually take her to Berlin and a veterinary degree in 1933. She managed to do this without her mother's knowledge, let alone support.

Von Maltzan was unusually politically progressive for a countess and was horrified when she read Hitler's *Mein Kampf*. As a result, she

* This from FemBio, "Maria Gräfin von Maltzan," https://fembio.org/english/biography.php/woman/biography/maria-graefin-von-maltzan/. Von Maltzan's autobiography, *Schlage die Trommel und fürchte dich nicht* (Ullstein, 1986) has unfortunately not been translated into English. The title means "beat the drums and don't be afraid."

volunteered with a Catholic resistance group who were helping Jews flee Germany, and she was part of the Solf Circle for a while, although fortunately no longer so by the time this resistance group was broken up and its members executed in 1943. She also worked with a Swedish church in Berlin who smuggled Jews out of Germany in secret compartments built into furniture that Swedish citizens were allowed to send home. She managed to remain above suspicion in part because one of her sisters, Alix, was married to Field Marshal Walther von Reichenau.*

Her closest call came when she hid the Jewish writer Hans Hirschel in her apartment. The Gestapo had been tipped off by a neighbour. They searched her place thoroughly, eventually becoming suspicious of the massive mahogany sofa-bed. They demanded that she open it, but she claimed that it was stuck. They were unable to figure out the mechanism, so they threatened to shoot it full of holes. She shrugged and stated that in that case they'd have to pay for the reupholstery. The Gestapo relented and left. Hans Hirschel was indeed hiding in a hollowed-out compartment in the sofa bed, presumably holding his breath. Hans and Maria married after the war. There are many more stories to tell from her time in the resistance, but they begin to stray too far from what I can justify in a brief history of veterinary medicine.

So, what of her veterinary practice? She had an exceptionally varied career and belongs right up there with her colourful predecessors, the Renaissance men and women of the profession. After the war, she worked as a circus vet and a relief vet, but she was bedevilled with health issues, which led to a drug dependency and several bouts of rehab. At one point she lost her licence and had to scrub floors for a living. However, the Jewish community never forgot her and helped her financially. Eventually she got back on her feet, and in one last remarkable episode, in 1975, at the age of 66, she opened a clinic in the Kreuzberg district of Berlin. Kreuzberg was in the shadow of the Berlin Wall and was the poorest part of the city. There she lived with a pet monkey and provided free veterinary care for

* Who was a war criminal whose troops cooperated in the Babi Yar massacre of almost 34,000 Jewish civilians.

the punks* who lived in the nearby squats. The quirky old woman and the radical young people got along very well. In 1987 von Maltzan was named Righteous Among the Nations by the Israeli Government, the highest honour they can bestow upon non-Jews. She died in 1997.

Von Maltzan's life and career spanned several epochal changes in veterinary medicine, so we're going to have to wind the clock back to 1919 now and catch up on some of these events. We'll begin by visiting with a Canadian veterinarian and his famous patient.

* I don't mean this pejoratively, but literally — West Berlin was one of the great centres of the punk rock movement, along with London and New York.

CHAPTER NINETEEN

Colebourn

"Do you know that it's been five years, three months, and seven days?" Captain Colebourn said, as he shelled a peanut he had fished out of a paper bag on his lap. "August 24, 1914, White River, Ontario. That's way back across the pond in Canada, if you'll recall. I remember it like it was yesterday. A beautiful late summer day. First hints of the coming autumn. Cool in the morning. The sumac going red already. But no hints of the type of war we were being sent to. None. We had no idea. We were in a jolly mood, me, and the rest of the boys of the 34th Fort Garry Horse." He paused and shelled another peanut. "Most of them are dead now," he added quietly. "Just a lark. Go to Quebec, then England, then France. Home by Christmas . . ."

He fell silent and ate a couple more peanuts, staring off into the middle distance. It was a typical leaden November day in London. It was objectively warmer than Winnipeg, but subjectively it felt colder here with the damp and the grey. Colebourn shivered and pulled his jacket tighter around his shoulders. "But anyway," he began again, his tone brighter, "August in White River. Long before we saw any action. There you were. Black as coal. Size of a cocker spaniel. Local fellow had you tied to the bench on the train station platform with a piece of twine. He was a trapper. He had killed your mother. And your twin brother had died somehow too. Seemed like a decent chap, though. Wanted you to have a chance. And wanted a few bucks for himself, I suppose." Colebourn laughed.

At this the bear finally reacted. She had greeted him when he arrived by coming up to the enclosure and grunting happily and then she had sat back down, attentively, but impassively, watching as the man spoke. When he laughed, she ambled back over to him and grunted again. Normally Colebourn would go right into the enclosure, which was an open terrace design, but today was different. He had come to visit much earlier than usual, before public hours, and he was sitting on the bench in his dress uniform. And, more significantly, there was an air of emotional heaviness about him. The bear sensed the difference and sat down again, watching her friend carefully.

"And do you know what I paid for you? Any idea?" he asked. "Twenty dollars. That's right, 20 whole smackers. A lot of money for a young veterinary surgeon, let me tell you. But I could see that there was something special about you." He paused again and cleared his throat. He had rehearsed this moment over and over. He was not going to upset her by crying. He had to pull himself together. "Winnie, I'm sorry, but this is goodbye. The real goodbye. Not the 'goodbye and see you soon' goodbye, but the 'goodbye and probably never see you again' goodbye."

Winnie licked her paw and sniffed the air.

"You've been very well looked after here, and they've promised to keep looking after you just as well. Do you remember the ocean voyage here? You did not enjoy that! Not one little bit. Going back would be even worse now that you're so much bigger. And you've become quite the star here. You are loved by so many Londoners. The zoo was very happy to accept my donation. As of tomorrow, you are officially no longer my bear. You will belong to the people of London."

Winnie grunted. Colebourn wondered how much she understood. She was a smart bear and knew quite a few words, but more likely she was thinking about breakfast. It was feeding time soon. Winnie looked around and sniffed the air again.

"Ha! You're definitely thinking about breakfast!" Colebourn said. "Well, maybe I'll stay for that, but then I must be off."

Doctor Harry Colebourn might not have merited a chapter in this book solely based on his connection to the world's most famous bear and one of the most beloved characters in children's literature, but he has two advantages beyond this part of his story. The first is that the site of his practice is a short walk from my house, and I am not above selfishly promoting a local hero, and the second is that the trajectory of his career neatly illustrates the most significant crisis the veterinary profession has ever faced.

In 1905, at the age of 18, Colebourn immigrated to Canada from England. He planned to attend the Ontario Veterinary College (OVC) but needed to work for a few years first to raise the necessary tuition funds. He held various jobs, including selling fruit on the street in Toronto and working as a deckhand on Great Lakes steamers. He graduated from OVC in 1911 and accepted an appointment from the Department of Agriculture to work in its Health of Animals division out west, in Winnipeg.

He joined the military soon after arriving in Winnipeg, training as an officer in the reserves while carrying out his government veterinary duties. By the beginning of the war, he was a lieutenant, initially, as mentioned in the story, with the 34th Fort Garry Horse, and then the Canadian Army Veterinary Corps (CAVC). There is a saying that goes, "Generals are always prepared to fight the last war." The last war for the Canadian and British armies was the Boer War (1899–1902), during which cavalry was still reasonably useful. And as we have seen, through the previous century cavalry was viewed as essential, right from the Napoleonic Wars, through to the Crimean, American Civil, and Franco-Prussian Wars. Moreover, quite aside from mounted cavalry, horses were essential for transportation of men and materiel. Full motorization of most armies wouldn't happen until the 1930s. While cavalry was quickly demonstrated to be worthless in the style of warfare that developed on the Western Front, the transportation function remained crucial. Consequently, there was an enormous demand for veterinarians. Canadian veterinary students in their final year were allowed to skip their final exams if they agreed to enlist. Ultimately, the CAVC had 72 officers and 756 other ranks, who treated 24,000 horses. Dr. Lisa Cox, curator of the Barker Veterinary Museum at the Ontario Veterinary College, has estimated that CAVC veterinarians returned 80 percent of their patients to service. This is a remarkable success rate, given wartime conditions and the fact that antibiotics were still three decades away.

Military veterinarians not only treated horses, but also several other species. Some of these you can probably guess at, such as dogs and mules. But some are surprising. If you visit the Parliament buildings in Ottawa (Canada's capital), and go into the Peace Tower, you will see a moving memorial carved in 1927 into the archway above the entrance to Memorial Chamber.* It has beautiful depictions of, from left to right, a reindeer, a mule, a carrier pigeon, a horse, and a dog. Below these there is a smaller carving showing two canaries in a cage and three mice beside them. The inscription reads, "The tunneller's friends, the humble beasts that served and died." "Tunneller" referred to the men in the trenches. There are no references to veterinarians treating the more unusual species commemorated in the archway, but it is reasonable to assume that they were involved in their care to some extent at least. Incidentally, the canaries and mice were kept in the trenches to serve as early warnings for gas attacks. Slugs were also used for this purpose but unfortunately did not rate a depiction in the memorial.

And the reindeer? In 1918, the Canadian Siberian Expeditionary Force (CSEF) was sent to Vladivostok in the Russian Far East as part of a doomed Allied effort to bolster Czarist forces against the Bolsheviks. Apparently, they had reindeer with them. To support them, and the other animals, the No. 6 Mobile Veterinary Section of the CAVC was also dispatched to Siberia. Incidentally, the official photo of this group of veterinary officers also shows two dogs who appear to be mascots. The

* Every day here, when the carillon tolls 11 a.m., pages are turned in the eight books of remembrance displayed in the Memorial Chamber. They record the names of the 118,000 Canadian war dead.

CSEF never saw combat, so casualties, human and animal, were confined to accident and illness.

London also has a monument to animals that served in wars. It was unveiled by Princess Anne at the Brook Gate to Hyde Park in 2004 and is much larger and grander than the one in Ottawa. The London "Animals in War" monument features horses, dogs, mules, and carrier pigeons as well, but it also gives space to camels, elephants, oxen, cats, and even glow-worms. The latter were sometimes used by soldiers for light. The inscription on the memorial reads, "This monument is dedicated to all the animals that served and died alongside British and Allied forces in wars and campaigns throughout time." And a second, smaller inscription reads, "They had no choice." That second statement is perhaps the most moving aspect of this beautifully designed monument.

Let's have one final unusual war animal story before we get back to Harry Colebourn and his famous bear. This one is also from Canada.

If you've driven the long, empty Trans-Canada Highway through Saskatchewan, you might have noticed an unusual sign advertising the local museum in the town of Broadview. The sign depicts a goat wearing a vest with military insignia. This is Sergeant Bill, and his stuffed body is a feature attraction. In a story very similar to Harry and Winnie's, soldiers passing through Broadview on the train in 1914 saw Bill on the platform and persuaded the owner to allow them to take him with them to basic training camp in Ontario. The soldiers then smuggled the goat with them to England and on to France. How, exactly, one smuggles a goat onto a troop ship is entirely unclear, but in any case, Bill, now Sgt. Bill, became a beloved mascot for the 5th Infantry Battalion of the Canadian Expeditionary Force. And his exploits were legion. There was the time Sgt. Bill pushed three soldiers into a trench just seconds before a shell exploded where they had been standing. Goats have an excellent sense of hearing. And then there was the time when he cornered three enemy soldiers who had entered the trench. He was also court marshalled twice, once for eating a soldier's bedroll and another time he butted a soldier with his horns after he had been fed incorrectly. Yes, there's a food theme to Sgt. Bill's misdemeanours. He was wounded twice, but survived the war and was eventually returned to his original owner. It would be neat if it had been Harry Colebourn who treated his war wounds, but we have no record of that.

Unlike Bill, Winnie did not go to the front. After Harry bought her, he took her with him to training in Quebec, where she became a big hit with the soldiers and was consequently permitted to sail with them to England in September 1914. There, Winnie and Harry, and in fact all the Commonwealth forces that were arriving, were dispatched to the Salisbury Plain, near Stonehenge, for further training and preparation. There are many stories from that time of Winnie's hijinks in the camp. She was an especially keen climber and would often shimmy up the tent poles, but as she grew, this threatened the tents with collapse, so the soldiers erected a special climbing pole for her, which she made ample use of. She was consistently reported to be gentle, affectionate, and keen to interact with people, traits that would eventually lead to her indirect immortalization.

In the meantime, Harry, and the rest of the CAVC, were occupied with the army's horses. As mentioned before, about 24,000 horses served with the Canadian Expeditionary Force, so there was plenty of routine veterinary work before even a single animal was injured in the fighting. Then, in December, it was time to go to the Western Front. There was no question of Winnie accompanying the troops. Not that there weren't other unusual animal mascots. We've already talked about Sgt. Bill. And then there was Tirpitz the pig that sailed with the British Royal Navy after being rescued from a sunken German cruiser. I'll put her ultimate fate in a footnote,* which you may wish to skip if you're of a sensitive disposition.

Wojtek's story is happier and more relevant to Winnie's. Wojtek was a bear who was adopted by the Polish II Corps during the Second World War. The free Polish forces fought with the Allies in Italy during Germany's occupation of Poland. There Wojtek was trained to carry heavy ammunition boxes and artillery shells. Wojtek ended his days peacefully in the Edinburgh Zoo.

* After four years as an apparently beloved mascot, Tirpitz was auctioned off for pork to raise money for the Red Cross. Her head remains mounted on the wall of the Imperial War Museum.

But unlike Wojtek, Winnie was no warrior bear. She had no place in the trenches, and Harry could only picture her being terrified there, or worse. No, she would need to find a home in England. So, Harry borrowed a motorcar, bundled Winnie — now the size of a small person — into the passenger seat, and drove to London. Just picture a soldier and a bear driving on an English country road. People must have been startled and amused.

And the amusement never ended. The story of how this bear cub from White River, Ontario, became part of the inspiration for Winnie the Pooh is well known and outside the scope of our narrative here, so we'll leave Winnie on her way to the zoo and return to Harry's story. He is the veterinarian, after all, and this is a book about veterinarians.

Harry Colebourn survived the First World War, although he was wounded in the head by a sniper's bullet and was affected by chlorine gas during the Second Battle of Ypres. He was awarded a number of medals and even recommended for the Order of the British Empire. Harry was evidently a good soldier and a good veterinarian. He returned to a changed country in 1919.

At the beginning of this story, I mentioned that during his life, veterinary medicine underwent its greatest crisis. Let's illustrate this crisis with a statistic. In 1900, there were 4,000 cars on the roads in the United States. In 1910, 500,000. In 1920, 7.5 million. In 1930, 23 million. You get the picture, and you know where I'm going with this. Most, if not all, these cars replaced horses. The dramatic upward curve for tractors was the same, but a decade or two behind personal cars, and the effect on the demand for horses was the same. It's an ironic turn of events given that one of the largest urban planning conundrums in the late 19th century was what to do with all the horse poop on the streets of large cities. In 1894, the *Times of London* predicted that at the current rate of growth in the number of horses, in 50 years the streets of London would be buried under nine feet of manure*. In 1898 an international conference to discuss this problem was convened in New York City. They didn't know what to do. But Henry Ford and Carl Benz did.

* Ben Johnson, "The Great Horse Manure Crisis of 1894," Historic UK, https://historic-uk.com/HistoryUK/HistoryofBritain/Great-Horse-Manure-Crisis-of-1894/.

The key fact to keep in mind is that when Harry Colebourn graduated in 1911, "veterinarian" was a fancy synonym for "horse doctor." As some of the previous stories have illustrated, veterinarians did treat other species, especially food animals and, though to a much lesser extent, dogs, but the great majority of them were occupied with horses the great majority of the time. The automobile revolution came on so quickly that the profession was thrown into a state of shock and panic. The closing of so many of the small private vet schools in the US was in part due to this. Doom was widely predicted, and the profession prepared itself to be drastically downsized. But, as you know, this didn't happen.

On return to Winnipeg, Harry set up a practice at 471 McMillan Avenue,* in the heart of the city. At that time there were still enough horses that he could focus on them, as that's where his experience lay. But already in 1920, he began seeing dogs and cats as well. And in this fact lay the seeds of the coming revolution. In the span of a generation, the image and reality of the veterinarian changed from being a horse doctor to a pet doctor. In 1952, when Norman Rockwell painted his famous depiction of a veterinary office, it was of a boy with a sick dog.

In 1926, because of poor health related to the chlorine gas exposure, Harry went back to his pre-war government job, but clearly his heart was in private practice because when he retired from his government position in 1945, he opened another clinic, out of his home at 600 Corydon Avenue.** Notably, this was an exclusively small animal practice, and he became known for his willingness to provide pro bono service to disadvantaged pet owners. Tragically, he died in an accidental fall two years later.

Unfortunately, the statistics on pet ownership do not reach far enough back for us to be able to plot those against the known historical trends in horse and car ownership, but even since 1988, there has been an increase in the number of American households with pets from 56 to 70 percent.***

* The clinic is long gone. There's a welding shop on the site now.

** Still there as a private home.

*** Terry L. Clower and Tonya E. Thornton's report, *Health Care Cost Savings of Pet Ownership*, which was prepared for the Human Animal Bond Research Institute. https://habri.org/assets/uploads/Health-Care-Cost-Savings-Report.pdf.

Of course, there is no direct correlation between these trends. Switching from a horse to a car doesn't create the need to buy a beagle. It's just a happy coincidence for the profession. Although perhaps there's an indirect correlation. Proponents of the concept of "biophilia" will state that there is an innate human need to be connected to the natural world. I agree with this view. From that perspective, it could be argued that the disappearance of horses from our cities, and our lives, created a void that people unconsciously craved to fill. Wolves on the hearth rug. Tigers on our laps.

Food animal work was always there, though, and remained steady through this period of upheaval; however, it never employed large numbers of veterinarians, and most of those vets needed horse work, and then later pet work, to keep their practices viable. This then brings us to the only veterinarian most people have heard of (other than their own, if they have one). Although primarily a food animal veterinarian, he famously treated "all creatures."

CHAPTER TWENTY

Wight

"Can ye come right away, Mr. Wight? One of our best pigs is doing right poorly."

Alf Wight glanced at his watch. 6:01. He stifled a sigh. Reggie had probably waited until what he considered to be a reasonable hour to call. Alf had a busy morning of TB testing planned, and he'd promised to try to have breakfast with his son, who had been complaining that daddy was always out. But there was no saying no. His partner in the practice, Donald, had a busy day ahead of him too, and in any case, although the sun was almost up, it was still technically night, and Alf was on call.

"Of course. I'll be there right away."

Alf stepped out into the pre-dawn chill and glanced up at the windows of the rooms where Joan and Jim were sleeping before climbing into his Morris Seven. It sputtered to life, evidently still willing to give another day's service despite being derided as an antique by everyone other than Alf.

Reggie Thompson's farm was up over Sutton Bank, towards Cold Kirby. Alf smiled to himself. If he had to be called out this morning, this was the best place to be called out to. It was going to be a glorious morning.

He quickly climbed out of the vale as he left Thirsk and ascended Sutton Road. He met the rising sun as he rounded the big bend past Cragg House. It shot past him, painting the wild moorland with vivid greens, golds, and russets. He pulled over. The sick pig could wait five minutes. Alf leaned out the window of the car and watched the mist in the vale begin to glow as the sun marched steadily across, east to west. He took a deep breath, smiled, and put the Morris back into gear.

"She were right as rain yesterday, but this morning she won't get up and she's covered in spots." Reggie twisted his cap in his hands as he said this. They were standing in the stone barn, ankle-deep in straw. The other pigs, on the far side of the barn, were grunting in what sounded like a healthy and contented way. Reggie's dog, Maggie, was by his side, panting, looking a little anxious, probably because her master was so upset. A bright beam of early morning sun came through the barn door, illuminating the pig in question. Alf knew this pig. Her name was Peggy, and she was Reggie's prize sow. Peggy was lying on her side, breathing very rapidly. Her eyes were half-closed. But most importantly, from Alf's standpoint, her skin was covered in purplish-red, roughly diamond-shaped, blotches.

He knew what this was.

"Am I going to lose her, Mr. Wight? Or can you do something to save her?"

Alf paused before answering. He knew what Donald would say. Donald was the master of the "art of veterinary medicine." That's not to say that he wasn't also fully competent in the "science of veterinary medicine," it's just that the art of it was his special forte. In the five years they'd been in practice together, not a week went by without Donald counselling Alf not to ever say anything would get better. "Always paint a black picture," he said. "If you say it's going to be all right and it dies, you're sunk! If you say it is going to die and it does, well, you have been proved right — but if it lives, you're a hero!"* Managing expectations and under-promising but over-delivering was, of course, wise advice, but Donald took it too far. Alf had his own style.

* This quote from Donald Sinclair is taken verbatim from Jim Wight's wonderful *The Real James Herriot*, pages 173–174. This whole story is very loosely based on an anecdote in Jim's book.

"I can certainly try something. It's a new treatment. Peggy has erysipelas, which is a bacterial infection."

Reggie nodded, looking solemn. "I've lost pigs to this before," he said quietly.

"I know you have, but with this new penicillin, I think we might have a chance."

"Owis it given?"

"Just a single injection, Mr. Thompson. No drenches or big tablets or anything like that. One needle in her hindquarters, and, if luck is with us, she'll be back to herself by tomorrow."

Reggie shook his head slowly. "I trust you, Mr. Wight, but I'll believe it when I see. My old dad did right well with his pigs, and he didn't believe in any of the modern vit'nry treatments Mr. Sinclair were selling. But it's a new day, and she's my best pig, so I'm willing to try it out."

"I appreciate that," Alf said, and opened his bag to fish out a vial of penicillin, a needle, and a glass syringe. "Here we go, girl," he said softly to the pig as he injected deep into her haunches. This would be the first time he used the antibiotic. Donald was already a big fan, having successfully treated mastitis in one of Bill Walton's cows with it last week. But Alf was a little bit nervous. It felt like his first time treating a patient as a newly qualified veterinarian back in 1940 all over again. To Reggie he was all confidence and smiles, though. "If we've caught it in time, this will do the trick." Donald would have grimaced had he overheard this.

The next morning, again at 6:01, the phone rang. It was Reggie.

"I thought for sure she were gonna die, but I'm telling you, it's a miracle. This morning she's taking her feed and moving about!"

"I'm so glad to hear that, Mr. Thompson."

"I don't know whether it were the new medicine or what, but I thank you."

It was the new medicine. The antibiotic revolution had come to veterinary practice. But before we explore that, a word, or several, regarding

Alf Wight. Many of you will know who I am talking about. And those who didn't know in advance would have picked it up from the footnote regarding Donald Sinclair's quote. Alf Wight is much better known by his pen name, James Herriot.* And Donald Sinclair was the Siegfried Farnon from the books and shows. James Herriot is inarguably the most famous name in veterinary medicine. By far. His books have sold more than 60 million copies in 36 languages. The beloved 1978–1990 BBC series based on the stories drew in millions of viewers and made him a household name. And the current critically acclaimed reboot is introducing Herriot to a new generation. But I haven't included him as one of my 23 representative historical veterinarians not only because of his fame and iconic status, but also because, similarly to Harry Colebourn, his career straddled the transition between two eras of veterinary medicine. When Wight began in 1940, veterinary work was still dominated by horses and farm animals, and treatments were primarily still curious patent medicines that the veterinarians often mixed themselves. By the end of his career in 1989 (incidentally, the year before I graduated), veterinary practice had been utterly transformed. Not only were we now in a world of antibiotics, but also of a thousand other new treatments and procedures that were often an order of magnitude more effective than what had been on offer 50 years prior. And the cow, horse, sheep, and pigs now shared the spotlight with dogs, cats, and a small Noah's ark of other pets. In fact, by the time Wight retired, more than half of veterinarians were engaged exclusively in small animal practice, something almost unheard of when he'd graduated. Wight himself was a passionate dog lover and was integral to the expansion of the small animal work at the practice in Thirsk. His son, Jim, recalls Alf as being something of an expert in wrapping cats. He could take the fiercest snarling, swatting tom, and in a few lightning-fast moves have him trussed up in an old towel like an angry sausage.

* He was forced to use a pseudonym by the Royal College of Veterinary Surgeons, which considered being listed as the author of a book sold to the public to be a form of advertisement, which was forbidden. The name was apparently inspired by Jim Herriot, a Scottish football (soccer) goalie who played for Birmingham City at the time. Jim, who was still alive at the time Wight's books became bestsellers, must have some funny stories to tell regarding mistaken identity.

Alf Wight's life also illustrates other truths about the transition to modern veterinary medicine. While it is unfair, and possibly misleading, to generalize, I'll do it anyway and say that the image of the generation of veterinarians that came before Wight was of bluster and showmanship, and a kind of no-nonsense authoritarianism, whereas Wight is emblematic of the modern image of the veterinarian: modest, sensitive, and compassionate. There was a dark side to his life as well, which also echoes the experience of too many veterinarians today. Alf Wight was stricken by crippling depression. It will shock many James Herriot fans to learn that this cheerful country vet living a seemingly idyllic life suffered a severe nervous breakdown in 1960. He had been prone to bouts of melancholy for years, but this was different. Alf even seriously contemplated suicide. Ultimately, he underwent electroconvulsive therapy, which helped considerably, as did writing, which his wife, Joan, encouraged him to do. Perhaps even more shockingly, the larger-than-life Donald "Siegfried Farnon" Sinclair not only contemplated suicide, but went through with it. In 1995 the recently widowered Donald took an overdose. He was 84. Alf had died of prostate cancer four months previously.

A full exposition of the evolution of mental health problems among veterinarians is well beyond the remit of this book, but it's worth a brief segue as an ever-growing plague of depression and suicide has stalked the historical development of the profession through modern times. Tragically, it is as much a part of our story as the manifold breakthroughs and victories.

According to some statistics, veterinarians today are more prone to suicide than any other profession or occupation. Think about that for a moment. Does that fit with your image of the job and the people in it? I suspect not. But now that you know about Alf's and Donald's struggles, perhaps it no longer shocks you. If those two superficially jolly characters can be so stricken, then it must be common in the profession. There are two major explanatory factors. The first likely applies to Alf's and Donald's stories, and the second applies to the growth of these issues since that time into the crisis it is today.

The first factor is the type of person attracted to modern veterinary medicine. It's not clear that this description fully applied to Donald Sinclair, but Alf Wight definitely was the modest, sensitive, and compassionate person I described earlier as typical for the new breed of vet. And that personality type is more likely to internalize the woes presented to them. Alf took everything to heart. Every failure, everything that could have gone better, every criticism.

The second factor is the nature of the job itself. I'll summarize the problem with the statement that veterinary medicine involves the relentless demand for high-stakes decisions with limited information and resources available. To be sure, human physicians make far higher stakes decisions — although, let me tell you, it doesn't feel that way when someone is sobbing and begging you to save their pet — but typically they have far more information in the way of patient history and diagnostic tests, and essentially unlimited resources. And since Alf's time, the stakes, or at least the perceived stakes, have escalated dramatically.

And there is a third factor that comes into play in Donald's tragic death — access to the means. Veterinarians perform painless and smooth euthanasia on a daily basis. The drugs are right there in their dispensary. They know how easy and effective it is. I'll leave it at that.

The story at the beginning of the chapter was also chosen to highlight the antibiotic revolution. No single change in veterinary practice had as much impact as suddenly becoming what Alf Wight's clients described as, the "magic man wi't needle." A golden age of miracle cures was born. A synthetic sulfonamide antibiotic had been discovered in 1932 in Germany* and was widely available, including to veterinarians. It primarily came as a powder which had to be mixed with water and poured down the animal's throat. This form of treatment was known as a "drench." While it was an enormous advance in that it was the first drug that actually killed bacteria, it had the potential for serious side effects and, as a drench, it was not

* Penicillin had been discovered four years earlier by Alexander Fleming, but his work was ignored until 1939.

always practical. It did, however, save Nero's life in 1944. Nero was a lion in the Royal Circus that was stricken with streptococcal pneumonia, a previously untreatable disease.

Injectable penicillin, once it became available at the end of the war, was safer, more efficacious, and far more practical. Suddenly a huge range of previously untreatable fatal diseases became relatively trivial matters. Veterinarians armed with the new "wonder drugs" were finally able to fully cloak themselves in the mantel of Science and leave behind most of the whiff of old-timey travelling medicine show hokum. As Alf's experience illustrates, even the crustiest Dales farmers were gradually persuaded that the veterinarian was more than an expensive nuisance of last resort.

What is of particular interest is that veterinary medicine was finally keeping pace with human medicine. After a more than a century of playing catch-up, innovations, such as penicillin, designed for human patients were given immediate application on the veterinary side. Today there is very little that is available for human patients that is not also available for veterinary patients.

And, in fact, in some cases the roles are even reversed. For example, vaccination for Lyme disease has been commonplace in dogs for more than a decade but is still under study in humans. Dr. Barbara Natterson-Horowitz, a human cardiologist, co-wrote *Zoobiquity: The Astonishing Connection between Human and Animal Health* in 2013 to highlight how much the human profession could learn from the veterinary profession, if only it could be bothered to pay attention. Her own epiphany came when she heard about a type of stress-induced heart failure that had been in the veterinary literature for years, but only relatively recently described in humans.

But to return to Alf Wight, a.k.a. James Herriot, for a moment, there is one more aspect of his life worth touching on here. Far more than anyone else, he inspired people to consider veterinary medicine as a career. It's impossible to overstate his impact on the image of the profession. Granted, the basic premise of the profession — helping heal animals — was already hugely appealing to many people, but presenting this with warmth and charm and humour was a winning formula. So many veterinarians who are moving the profession forward today were originally inspired by Alf. We all should be grateful to him.

I mentioned that Alf was a dog lover, and we've met a few others before him. It's time we met a cat lover.

CHAPTER TWENTY-ONE

Camuti

"I would like to speak to Dr. Louis Camuti, please." The man on the phone sounded businesslike but cautious.

"You're speaking to him. How can I help you?"

"I understand that you," the man paused, as if carefully weighing his next words, "specialize in treating cats."

"That is correct."

"And that you do house calls."

"That is also correct."

The line crackled for a moment with neither man speaking.

Louis cleared his throat. "I take it you have a sick cat you'd like me to see?"

"My client does."

"Your client?"

"Dr. Camuti, my client is Mrs. de Havilland. Mrs. Olivia de Havilland." Camuti groaned inwardly. A movie star. He hoped she wasn't anything like Tallulah Bankhead, who was impossibly obnoxious and demanding. But he had enjoyed de Havilland in *Gone with the Wind*, a film he otherwise found dull and overlong, so he supposed he should give her the benefit of the doubt.

"And she has a sick cat that she'd like me to see?"

"Yes." The man cleared his throat. "Ding Dong* is under the weather."

"Under the weather? So not that urgent?"

"Oh, no, Doctor. I assure you it is urgent. Mrs. de Havilland is extremely concerned about him. She will not be able to go on stage tonight until she is confident that he is going to recover."

Camuti had heard that she was in a new production of *Romeo and Juliet* on Broadway. He had also heard the whispers that she was too old for the role. But none of that was relevant. So long as she stayed out of his way and didn't bark orders at him like Bankhead did, she could be the Queen of England for all he cared.

"Let me check my agenda," Camuti said. He didn't have to, but he didn't want this fellow to believe that he could just be commanded to come. He knew that his next appointment, a tubby tabby named Frederick, had cancelled after Mrs. Browning discovered that he wasn't constipated after all but rather had been stealthily pooping behind the divan. "Ding Dong is in luck. I have a cancellation and can be there in half an hour, traffic permitting. Tell Mrs. de Havilland to lock Ding Dong in the bathroom."

Camuti called his wife, Alex, who did all the driving. Parking was such an enormous headache that he needed a driver to be able to stay on schedule. Alex would circle the block, double-park, or park illegally, as needed. Meter maids had come to call her the "fire hydrant girl" after her favourite place to wait for her husband while he attended to his feline patients.

The doorman knew to expect Camuti and let him in immediately. Camuti was long since past being impressed by fancy lobbies, gilded elevators, and liveried servants. And he was preoccupied as he ascended to the apartment with considering what might be ailing Ding Dong.

---

* Despite my best efforts, I was unable to find the names of any of Olivia de Havilland's cats, so Ding Dong is a pseudonym. De Havilland was, however, one of Hollywood's greatest cat lovers and a particular devotee of the Siamese breed. Although she lived in LA and Paris for most of her remarkable 104 years, she was in New York from 1950 to 1953, where Camuti became the attending veterinarian to her cats.

Mrs. de Havilland's agent had gone on to describe the symptoms as an absence of his usual "joie de vivre." Ding Dong was also apparently 20 years old. Camuti hoped he wouldn't have to put the cat to sleep. After all these years, that was still the procedure he dreaded the most. It probably didn't help that his very first patient as a newly minted veterinarian had been a St. Bernard named Tiny that he was supposed to euthanize. He had made the mistake of attempting to do this in a small unventilated bathroom, so the chloroform knocked him out before it had any effect on Tiny.

"Please come in, Doctor." Camuti was mildly surprised that Mrs. de Havilland met him herself at her door. He was expecting the agent, or a butler. She was even more beautiful in real life than she was on film.

"Thank you. Is the patient secure in the bathroom?"

"I'm so sorry, Doctor, but I couldn't find the little devil!" Despite her many years living in America, she had a hint of an English accent. "It's as if he knew you were coming. He's never liked doctors."

Camuti nodded. This was annoying, but not entirely unexpected. At least she had apologized. "Then we will have to deploy the Camuti Method."

"The Camuti Method?"

"This is a common problem in my practice. I can't afford to stand around until the errant pussycat turns up, so I have developed a foolproof method of finding them. We open all the doors and then we begin at the far end of the apartment and go through room by room, closing the doors one by one behind us for each room we have searched. Searching under furniture and behind furniture is key. They will dart this way and that, but eventually they have nowhere left to run."

The actress nodded. "Cecily and Malcolm will help us."

And so the animal doctor, the actress, her maid, and her agent set about scouring the apartment, room by room, Malcolm glancing anxiously at his watch from time to time. Two other Siamese cats — Ralphie and Miss Primrose — were located, but Ding Dong somehow remained elusive. It were as if he had learned the trick of invisibility.

"Oh dear, where can he be?" de Havilland said, looking to be on the verge of tears. It seemed the Camuti Method had failed. But Camuti just grinned.

"Let's go back to the master bedroom. Make sure you close the door behind us again. I have an idea."

As they walked back down the hall, Camuti smiled to himself, thinking about how he always got a kick out of blandly slipping the statement "I've been in a lot of movie stars' bedrooms" into conversation at Manhattan cocktail parties. Fortunately, Alex was a good sport and always laughed along with the joke.

"You have a box spring under your mattress, Mrs. de Havilland?" he said as they gathered at the foot of her exquisite antique four-poster bed.

"Yes, I do." The actress looked puzzled. Malcolm looked at his watch again.

"I apologize to you and to Miss Cecily, but we're going to have to mess up your well-made bed. First, we will take the mattress and all the covers and sheets off, and then we will have a careful look at the box spring."

Once this was done, Camuti stalked around the perimeter of the box spring, occasionally poking at the sides, sometimes bending down to poke at it from underneath. After a couple of minutes he stood up straight and proclaimed, "Aha! Just as I suspected." He pointed at a spot just under the edge of the box spring and then inserted his fingers into a tear he had discovered there.

"Now, each of us will take a corner and pick it up. Then we will gently tilt it back and forth."

Wordlessly, the actress, agent, and maid did as they were told.

A low growl issued from deep inside the box spring.

"Malcolm, please lift your corner as high as you can." Malcolm was at the farthest corner from the tear. Fortunately, he was also the tallest of the four.

The growl turned into a yowl and then a howl.

"Oh, my poor baby!" Mrs. de Havilland cried. "Don't worry, the doctor is here to help you! I promise!"

"Give it a gentle shake," Camuti instructed.

A moment later a long, thin, disgruntled-looking elderly Siamese cat emerged from the tear in the box spring fabric. Mrs. de Havilland scooped him up and hugged him to her while he struggled and meowed pitifully.

Camuti smiled at them. "Your dressing table will do nicely as an examination table. Please set him down there, but keep a hold on him."

The actress did as she was told, and Camuti proceeded to examine the still-protesting cat. After a few minutes he stood back and smiled at Mrs. de Havilland. "Ding Dong will be fine, madam. He is, as you know, old. And that is all that is the matter with him. The elderly have more off days than the young. This is simply one of them. I have just the thing for the old boy. I will give him an injection of a tonic that will put some pep back into his step."

"Oh, thank you so much, Doctor. I thought he was dying! I couldn't bear the thought of that. I am ever so much in your debt. Malcolm, please see that Dr. Camuti gets two of the best tickets in the house for tonight." She picked up Ding Dong again after the injection, who immediately squirmed out of her arms and bolted for the far corner, where he managed to squeeze himself under a dresser.

Louis Camuti was the first veterinarian in the world to specialize in cats. He started his cat practice in Manhattan in 1933, which is astonishing. There's no record of who the second cat vet in the world was, but I'm willing to wager that they started at least a decade or two later. And Louis stayed in practice for 60 years, which is also astonishing.* Originally, he had a clinic at 1020 Park Avenue, but by the mid-1940s he transitioned to primarily performing house calls. Picture yourself trying to carry your cat to see the vet down a busy New York sidewalk or bringing it in the subway or taxi, and you'll understand why house calls were popular with his clients. Louis felt that many cats were missing out on vital health care because it was too difficult to bring them in. He also thought that it was better to observe them in their home environment, although the cats themselves often didn't see it that way. The title of his entertaining

* That's more than double the average career length, but it's not a record. As of this writing in 2023, Dr. Paul Wise, in Chico, California, was still seeing patients two days a week. He graduated in 1950. He's 104 years old. True story.

memoir, *All My Patients Are under the Bed*, says it all. A lot of time was spent looking for the patient and then, once located, trying to extract it from its hiding place. Hiding in box spring mattresses was just one of the tricks his patients got up to (although I'll confess that there is no record of that specifically happening with Olivia de Havilland's cats). Recalling the number of clients he saw on all fours, coaxing cats from under beds, Louis joked that he knew the hind end of some clients better than the front end.

The concept of a cat practice was remarkably progressive, but of course his medical approach was a product of the times. Consequently, there are toe-curling stories of him spaying cats on people's kitchen tables. Crazy talk now, but not in the '40s and '50s. The "tonic" he used on Ding Dong was a proprietary mix of hormones and vitamins, the specific composition of which he appears to have kept to himself. Although it is no longer accepted practice, up until the late 1990s we were still routinely using anabolic steroids — the kind banned in professional sports — to perk up elderly cats, and B vitamin injections still have several valid indications today.

As the story indicated, Louis Camuti had many celebrity clients. He eventually became quite the celebrity himself, even appearing on *The Tonight Show Starring Johnny Carson* in 1962. Camuti died of a heart attack in 1981, at the age of 89. He was on his way to see a patient.

Besides being entertaining, Camuti's story has a place here for three interrelated reasons. The first is that as a pioneer in feline medicine, he represents another significant step forward in the profession's evolution from being entirely practically oriented towards one where emotion plays an ever-larger role. In fact, his career was exclusively based on emotional reasons for providing medical care. Although some cats were kept as mousers, the great majority were purely companions, more so even than dogs, a much larger percentage of which were "useful," in such ways as for herding, guarding, hunting, or other sport. In his autobiography, Camuti explains that he stumbled into his feline specialty. Nobody expected that there would be such a demand to treat cats in the 1930s. This was a time when they were generally considered to be more or less disposable pets. But when word got out that he was willing to do house calls for cats, the demand for this service quickly displaced all other aspects of his practice. The minority who cherished their cats was more than large enough, at least in New York.

All My Patients Are under the Bed is full of stories exemplifying this cat love. There's the couple who planned to fly him to Italy to attend to their ailing cat who was there on vacation with them (the cat got better before this could be arranged). And there's the woman who ran afoul of the FBI for trying to import black-market fish during the Second World War for her apparently extremely fussy cat. Only Japanese tuna would do. This kind of devotion was far more widespread than the culture openly acknowledged. That's always the way. What people do in private is always several steps ahead of what is openly considered acceptable.

The second reason is that this is obviously also an example of the broadening species base for the profession. For centuries veterinarians and their kind focused on horses and cattle, with the occasional glance at other livestock and, even more occasionally, at useful dogs. From the mid-20th century onwards, cats became integral to veterinary practice, now accounting for up to half of patients in many small animal clinics. And in the last 50 years, the number of species we routinely see has continued to expand. I have personally treated somewhere around 20 species, and I am not even an exotic pet specialist. While some of these animals were children's or ornamental pets with looser emotional bonds, the degree of love and connection people can feel for any animal, regardless of species, should not be underestimated. The short-lived pets present problems in this regard, but the ones with longer life spans such as rabbits and many birds are now routinely described as family members.

And the final reason Camuti's story is illuminating is that he was essentially a specialist, and specialization is an ever-growing trend in veterinary medicine. He was not formally trained as a specialist. That training didn't exist yet, but the second half of the 20th century would see an explosion in these training programs and the subsequent accreditation of specialists.

In the United States, the veterinary pathologists were the first out of the gate to organize themselves, forming the American College of Veterinary Pathologists in 1951. This was followed by the veterinary ophthalmologists in 1957, the surgical specialists in 1965, the radiologists in 1966, the internal medicine specialists in 1973, the dermatologists in 1974, and the dentists in 1987. (My apologies if I've missed your favourite specialty.) As of 2020, there were 13,539 board-certified specialists in the US, out of a total of 118,624 veterinarians. When Louis Camuti started, there were zero board-certified specialists. Incidentally, there are 75 certified specialists in feline medicine in America today, and the American Association of Feline Practitioners (AAFP) has more than 4,200 members.*

Wight and Camuti bring us into the reach of present-day memory. People are still alive whose animals were treated by them, as are veterinarians who were their freshly graduated young colleagues. Their careers saw the beginning of the explosion in small animal medicine that continues today. Long paragraphs, even chapters, could be written laden with statistics illustrating this phenomenon, but instead let's look at a specific story, drawn from my own practice.

When my first boss, Dr. Elmer Clark, opened Birchwood Animal Hospital in Winnipeg in 1959, it was only the second clinic in the city to exclusively see pets. Two clinics, one vet each. The population at the time was 450,000. Metro Winnipeg now has 840,000 people and 50 small animal practices employing well over 100 veterinarians. Clark tells the story that the clinic was almost instantly busy. There was no need to advertise or worry about meeting payroll and expenses. You unlocked your

* AAFP members are not true specialists. The distinction may perhaps seem technical to the outside observer, but it is an important one. Becoming a board-certified specialist in a species is much more unusual than becoming board certified in a system specialty, such as cardiology or dermatology. Most veterinarians who decide to focus exclusively on cats simply do so on their own without extra training. They often join the AAFP for support and access to resource materials and conventions. This is much easier and less time consuming than the multi-year training involved in becoming a formally recognized "specialist." The same goes for cow vets, chicken vets, etc.

door in the morning, and people would stream in with their pets. Just like Camuti experienced, there was tremendous pent-up demand. The recent COVID-19 pandemic led to a surge in pet ownership which has recreated this phenomenon. None of us in small animal medicine can keep up.

For large animal veterinarians, the trend has been the opposite, despite rising numbers of livestock. To cite just one example, in Canada the number of pigs quadrupled from 1921 to 2011. Yet, the number of veterinarians attending to those pigs dropped significantly.* Far more pigs per vet. The same goes for cattle, sheep, chickens, you name it. The reason is, of course, the intensification of agriculture, leading to many more animals per farm and a drive for efficiency to keep medical expenses to a minimum. As a result, the modern food animal veterinarian mostly deals with problems at the herd level rather than the individual level. Prices fluctuate wildly, but the recent average value of an individual pig in Canada has been around $200. This happens to be pretty close to the cost of the average veterinary visit for a pet. It's easy to see why food animal medicine has been forced to take a different path. We'll visit with one of the pioneers in herd health medicine in the next chapter.

Consequently, with the exception of veterinarians in the shrinking "James Herriot niche" of working with small family and hobby farms, the profession is split. In many ways, those of us practising exclusively small animal medicine now have more in common with human medical doctors than with our colleagues practising food animal medicine. This has increased talk of formally separating the profession, although the subject is still taboo in many quarters. Veterinary students continually grumble about "wasting time" on learning about species they never intend to treat. And the fact that someone can practise exclusively on cats for 30 years and still have a valid licence to treat a horse does raise some questions. In case it isn't clear, cats are as medically different from horses as you are from either species. And 30 years is long enough ago that it may as well

* In the US, according to the American Veterinary Medical Association's 2022 report, only 3.9 percent of veterinarians exclusively treat food animals. Another 4.2 percent are in mixed practice, so it's only 8.1 percent that would *possibly* see a pig. It's safe to say that in 1921 a solid, if not overwhelming, majority of veterinarians would have been able and willing to treat pigs.

have been never. However, in many remote or small communities, there is still a demand for the vet who sees "all creatures," so the talk is just talk. For now.

For our next-to-last stop, we're going to move ahead to the early 1980s and go literally halfway around the world from Louis Camuti's Manhattan.

CHAPTER TWENTY-TWO

Blood

Tim looked around the office while Dr. Blood tried to wrap up his phone call. The old prof had gestured for him to sit and then rolled his eyes and pointed at the receiver with a mock grimace to indicate that it was a boring conversation.

Something to do with committee work, it seemed.

Tim had been in here a couple of times before. The view across the Werribee Campus of the University of Melbourne was nice, but he particularly liked the chalkboard covered with half-erased numbers, the overstuffed bookcases, and the various framed photographs and diplomas on the wall. It was all very . . . professorial. His eyes eventually settled on an old picture of a Holstein cow.

"Ah, yes, she's a beaut, isn't she?" Blood said, noting Tim's interest as he hung up the phone. "Her name was Esmeralda. Gift from a grateful client back when I worked in Canada."

"A beaut, for sure. Makes me think of your lectures on the examination of the ox." Tim emphasized the word "ox" because of how comically old-fashioned it sounded.

Blood chuckled. "You remember those?"

"Of course! First the distant exam, looking at the environment, and how the animal holds herself. Walk all around. Then approach the nose and begin the close examination from one end to the other. Use all your five senses." Tim laughed.

"Ha! Very good! But just five, Tim?"

"Pardon me, sir. Six senses. But the sixth is not intuition. It's experience and learning from your past errors."

"Not exactly a sense, but yes, that's correct. That's what I taught. I'm impressed that you remember, given everything we try to cram into you poor students." Blood chuckled again while cleaning his glasses. He held them up to the light and, satisfied, put them back on. "But I didn't ask you to pop by in order to reminisce about the past. It's the future I want to chat about."

"The future?"

Blood nodded and pointed to a cardboard box in the corner. It had the IBM logo on it. "That is a brand-new IBM 5150 microcomputer."

"Cool."

"I hope so! I've been told that you know something about microcomputers. Is that true?"

"A bit. I'm in the campus computer club, and we have an Apple II at home. My dad's an engineer."

"Good enough. And I've also been told that you're looking for a summer job."

"Yes, I am. I was thinking in a lab on campus somewhere, but anything really."

"If you're interested, I might have something for you here. For a number of years now we've been asking farmers to fill out data sheets regarding various reproductive and health parameters and then mail them in. We feed these into the university's mainframe computer. The computer analyzes the data, and we look for trends that we can use to give the farmer advice. But this is cumbersome, slow, and costly. You can imagine that, right?"

Tim nodded, picturing those old giant computers with whirring reels of tape being fed punch cards.

"So, my plan is to develop a system whereby an ordinary veterinarian in private practice can use a microcomputer to run software for his dairy clients. More direct, cheaper, and timelier. A 16-bit machine like this with 128k of RAM and two drives for 1.2 megabyte floppies should be able to handle the data for anywhere from 50 to 100 farms with an average of 120 milking cows each."

"Wow, those are quite the specs," Tim said.

"They are, and it is out of reach of the average farmer, but a busy practitioner should be able to justify it. Moreover, he could also use it to manage the practice financial records, do some word processing, store drug dosage charts, or even keep track of bird sightings!"*

Tim raised his eyebrows fractionally at the latter suggestion. "And the faster machines are only going to get cheaper and cheaper!"

"Exactly. Now, Tim," Blood leaned forward, "what I'm hoping you can help me with is to write a mastitis** management program. There are reasonable options out there already for reproductive and nutritional management, but I feel it would be very helpful to have one that would analyze milk cell counts. It could print out lists of cows suspected to have subclinical mastitis."

Tim chewed his lower lip. On the one hand, this was a fantastic opportunity to work with the world's leading bovine medicine specialist, but on the other hand, his coding experience was limited to a silly text-based Monty Python game he and his mates had written using Basic.

"I'd ask one of the fellows from the Computer Science department," Blood went on. "But I think some veterinary background would be helpful. At least in the initial set-up. We can get the eggheads over there to fine-tune it if we need." He chuckled and smiled broadly at the student.

What the heck, Tim thought. He smiled back. "You bet, sir, thank you for the opportunity!"

---

* Blood was an avid birder. He actually made this specific suggestion in a paper he presented at a 1984 cattle disease congress in Durban, South Africa.

** Inflammation of the udder. This is a common affliction in dairy cattle that has a significant impact on milk output. The subclinical variety, without obvious symptoms, is particularly insidious.

Tim is fictional, but Dr. Douglas Blood definitely is not. Born in England in 1920, he immigrated to Australia with his parents when he was young. He received his veterinary degree from the University of Sydney in 1942 and, after a stint in the army, went on to teach at Sydney; Cornell, in Upstate New York; and Guelph, in Ontario, before returning to Australia in 1962, where he took up a position at the University of Melbourne. He retired in 1985, the year, incidentally, that I began my own veterinary studies. As it happens, my large animal medicine professor at the Western College of Veterinary Medicine (University of Saskatchewan, in Saskatoon) was Dr. Otto Radostits, who co-authored several editions with Blood of the seminal *Veterinary Medicine: A Textbook of the Diseases of Cattle, Sheep, Pigs and Horses*.

As the story highlights, Blood was one of the key figures in modernizing the concept of herd health in veterinary medicine. In Alf Wight's time, farm animals were generally dealt with on an individual basis and also with what is sometimes called "fire engine medicine." This involves the veterinarian darting around the countryside "putting out fires" — in other words, dealing with random acute illnesses and injuries. By the middle of the 20th century, with the growing intensification of agriculture described in the last chapter, this was becoming a less and less viable approach. The alternative was to look at the entire herd, or flock, as a unit, and to focus more on prevention through vaccinations, environmental management, nutritional science, and careful surveillance, including data analysis. Blood was at the forefront of the application of computers to herd health management. Yet, he was also a consummate diagnostician of individual animals. His thorough and comprehensive "six senses" approach (don't ask about taste . . .) to the physical examination process was subsequently taught worldwide. That this was necessarily slow and low-tech makes it an interesting counterpoint to the speed and high-tech of the computerized databases he also promoted. There was, and is, a need for both in veterinary medicine. Today we are still encouraged to take a "high-touch, high-tech" approach to practice.

Computerization also made rapid inroads into small animal practice in the late 20th century, but for invoicing, inventory management, and eventually medical records keeping, rather than for data analysis. However, although it would be quirky to remain paper-based and analogue, computerization is essentially optional in smaller pet clinics, whereas it is mandatory for modern herd health–focused practices. It is impossible to imagine the three hundred members of the American College of Poultry Veterinarians dealing with the 1.5 billion chickens (plus 200 million turkeys and 30 million ducks) in the US without substantial information technology support.

In the 40 years since Blood's time, the profession has evolved at an ever-quickening pace. Food animal medicine's focus on technologically sophisticated herd (and flock) health management has continued to intensify, while the companion animal side of the profession draws closer and closer to the standards and practices of human medicine. In parallel to these trends, veterinary medicine has also become much more female, more diverse, more global, and much broader in scope. With that in mind, let's move ahead to this century and into Bwindi Impenetrable National Park for our last story.

CHAPTER TWENTY-THREE

Kalema-Zikusoka

"I don't think Kacupira likes this, Gladys," Kock whispered, inclining his head towards a large black shape in the dense green foliage up the slope ahead of them. The two veterinarians crouched with their dart gun and medical supplies, trying to remain unobtrusive, although they had just announced their presence by darting Kasigazi.

"You are right, Richard. Kacupira knows that this is not a normal visit. He will want to protect Kasigazi. But I do not believe he will charge us," Kalema* said.

The big silverback mountain gorilla began walking down the slope towards them. Slowly, but deliberately.

Kock shot Kalema a glance.

She chuckled quietly. "Do not worry. Even if he charges, I have been charged before, and it has been fine. Stay calm. Make yourself small. Look away. Move backwards very slowly. That is what we normally do.

* This was a few years before Dr. Gladys Kalema married Lawrence Zikusoka and changed her name.

But so that we can do our work with Kasigazi, I will ask the trackers to scare him away."

Kalema quickly issued a number of orders. In seconds a line of men formed between the 350-pound gorilla and the veterinarians. They clapped loudly, stepping slowly towards Kacupira. The gorilla stopped and then ambled off to the side. No matter how much the trackers clapped, Kacupira remained within 15 metres (49 feet) of the veterinarians, but he was no longer trying to approach them. He was intent on watching what they were doing with the now very sedate juvenile.

Kalema suspected scabies. Moreover, she suspected that the gorillas had gotten it from contact with humans along the edge of Bwindi Impenetrable National Park.* It was common for the people to put up scarecrows and to dress them in old clothing that could be infested with the mange mites that cause scabies in both humans and gorillas. Because of how cold it could become at night, the resulting hair loss could be fatal in some gorillas, especially the small and young. Also, the itchiness was intense, resulting in constant painful scratching and the risk of secondary infections.

Kasigazi was the second-worst affected gorilla, with about one-quarter of his hair missing. Unfortunately, the worst affected one, an infant named Ruhara, who had lost three-quarters of his hair and was clearly suffering, could not be helped right away because his mother was shy. Kalema and Kock could not get near enough to dart her and therefore be able to examine and treat her baby.

"I think you're right, Gladys," Kock said as he carefully inspected the sleeping ape's skin. "As we all know from those derm lectures in vet school, there are a hundred different skin diseases and only about three ways they present, but I'm willing to bet this is scabies."

"He is so itchy," Kalema said, shaking her head. "He is still scratching himself as a reflex, even with the ketamine anaesthetic. He has so many

* "Bwindi" comes from the Runyakitara word for "a place full of darkness." It's called the "Impenetrable Forest" in English because its extreme density makes walking impossible without hacking out a path. The park is in Uganda, and the forest extends into the Democratic Republic of the Congo.

wounds. It is good that we have long-acting antibiotics for him, as well as the ivermectin for the scabies." She glanced over to Kacupira, who continued to stare at them intently. Although she knew that eye contact should be avoided during a potentially hostile encounter, the silverback showed no signs of aggression anymore, so she allowed hers to meet his for a second. His dark brown eyes were soulful and alive with intelligence. She hoped he understood that they were there to help.

The above story has been adapted from Dr. Gladys Kalema-Zikusoka's marvellous autobiography, *Walking with Gorillas: The Journey of an African Wildlife Vet.* I encourage you to pick up a copy if you have even the vaguest interest in Africa, conservation, development issues in the Global South, or, of course, wildlife medicine. Her life has been remarkable, and her resumé fairly bristles with firsts.* She was the first Ugandan wildlife veterinarian, the first female African wildlife veterinarian, and a "One Health" pioneer. In many ways, One Health is an ancient concept, intuitive to pre-industrial peoples, that recognizes that human health, animal health, and environmental health not only interact, but are also inextricably bound together. According to the U.S. Centers for Disease Control and Prevention, and the One Health Commission, the start of One Health as a formal modern program was with a 2004 Wildlife Conservation Society conference called One World, One Health. Neither mentions that a year prior, together with her husband, Kalema-Zikusoka founded the NGO Conservation Through Public Health (CTPH). CTPH is a quintessential One Health program. Her insight was that the health of the gorillas was linked to the health of the people living near gorilla habitats. Not only could diseases like scabies and tuberculosis be transmitted to the gorillas, but also healthier people would make happier neighbours for the gorillas and would ultimately be less likely to turn to poaching or habitat destruction.

* And it bristles even more with awards, accolades, and appointments. For the full impressive list — far too long to reproduce here — I recommend you check out her profile page on the National Geographic website (see Sources and Further Reading at the end of the book).

And she turned out to be right. It became a classic virtuous circle as CTPH's community-based approach to human health improved the lives of the people, which in turn improved the health of the gorillas, which boosted tourism, which brought more jobs and money into the community, improving human health further.

However, CTPH had a slow start with skepticism, racism, and sexism as significant barriers. But with diligence and Kalema-Zikusoka's deeply felt belief in the validity of this approach, it has succeeded beyond all expectations. In the last 20 years, the population of mountain gorillas in Bwindi has increased more than 40 percent, a remarkable achievement given the multiple existential threats this species faces. Moreover, the people in the surrounding communities are enjoying better health, better education, more employment opportunities, and an overall markedly better quality of life than before CTPH came into the picture. The program has now expanded beyond Bwindi to three other areas in Uganda.

Incidentally, in a coincidental callback to the origins of the profession, Kalema-Zikusoka has also worked on rinderpest. (If your memory is hazy, recall the pope's cows in chapter eight).

Kalema-Zikusoka is emblematic of the future of veterinary medicine. Not only is the One Health principle she champions destined to play an ever-larger role, but also her gender, nationality, and career decisions all throw open a window onto where the profession is headed.

We met some of the first female veterinarians in Elinor McGrath's day. They were the first small tremors of what would become a seismic shift in the profession by the late the 20th century. After McGrath, Reid, and Cust, the number of women in the profession hovered in the low single digit percents for decades. By 1960 it was still just 2 percent in the US. Then it began to rise in the early 1970s, at first gradually, and then dramatically by the late 1980s. Gender parity was achieved in most parts of the Western world in the early 2000s. Today about 63 percent of practising veterinarians in both Canada and the US are female. And about 80 percent of current veterinary students are women. The future is female.

The numbers are lower in other professions. In law, dentistry, and medicine, roughly half the students are female.* What makes veterinary medicine so different? Don't men love animals too? Of course they do. To answer this question, I'm going to quote myself with an excerpt from *How to Examine a Wolverine*:

> It's complicated. One factor is that competition to get a spot in veterinary school is ferocious, more ferocious than for any other profession, and young women are increasingly in a better position to win that competition. Women now dominate in academics, often occupying the top rungs in the lists of the best students in any given class. The reason for this is beyond the scope of this essay, but there are innumerable articles on the subject and if you read one, you'll see that the falling academic performance of young men is the cause of much handwringing.
>
> Another factor is that veterinary medicine pays less well and is perhaps less prestigious than many of the other professions. This is a terrible statement about the state of gender relations in our society, but women historically have accepted lower pay and men historically have been encouraged to seek prestige. These things are changing, but some ingrained cultural norms will take a long time to truly fade away.
>
> And finally, veterinary medicine requires more empathy than any other profession. If you are reading this, then you already know why this is true. Again, this is likely more a statement about our culture than anything else as I don't think I am any less empathetic than my female colleagues, but perhaps I am less concerned about those subtle cultural

* Engineering is an outlier in the other direction. Only around 20 percent of engineering students are women right now.

> signals. This is not to absolutely deny the influence of biology. In a survey of very anxious dogs, seven out of ten preferred a female veterinarian. Something about men's deeper voices and harder features freaks them out. (I'm joking, of course. To the researcher's endless frustration, the dogs were unable or unwilling to answer the questions. But the observation is generally true.)*

Wherever the reasons lie, veterinary medicine has been through a more profound and complete gender reversal than any other profession or occupation. If you're picturing a veterinarian in the future, or today for that matter, you should picture a woman.

Kalema-Zikusoka also represents the rise of the profession in the world outside of "the West." As has been discussed before, a connection to animals is a universal human trait. It is not specific to any one cultural group. Even where this connection is entirely practical because the animals are only valued for their food potential, rising incomes, standards, and expectations lead to an increased demand for veterinary services for these animals. And everywhere in the world, as people enter the middle class, they acquire companion animals. Most major centres from Lima to Lagos to Lahore have modern pet hospitals that would not look out of place in my neighbourhood. This is only going to increase, and quickly. I could proffer a blizzard of statistics to back up this assertion, but that would make for dull reading. Instead, be satisfied with the January 10, 2022, article in the *India Times* which stated that the pet population there was growing at 12 percent a year. Temporary pandemic bump aside, this is an order of magnitude higher than here. More pets means more vets. In many parts of the world, people with one- or two-year technical courses have historically taken on the role of veterinarian, but increasingly the demand is for university training, equivalent to that in the Western world.

And finally, Kalema-Zikusoka's story highlights the diverse career paths a veterinarian can take. Not only does she treat gorillas, but also

* Schott, *How to Examine a Wolverine*, 167–169.

giraffes, elephants, antelopes, and other African animals. And now she finds herself in a primarily administrative role. Veterinarians are sometimes uniquely qualified to run animal-oriented charities and enterprises. In my own graduating class, we have a salmon vet, a crustacean pathologist, a neurologist, a laboratory animal vet, a food safety specialist, a researcher, and a professor, in addition to all the horse, pet, and food animal vets.

At this point, since we're approaching the end of the story, I could lay out a timeline of all the great advances in veterinary medicine over the last half century. The first prosthetics. The first laparoscopic surgeries. The first laser procedures. The first MRIs. The first stem cell treatments. The first . . . You get the picture. If it's available for humans, it's probably available for animals somewhere. However, this would make for dull reading, and many of these headline-worthy breakthroughs are for a minority of patients, usually pets cared for by a privileged few. The bigger story is the expansion of basic care for animals worldwide. Many more vaccines. Many more simple, humane procedures. Many more effective and affordable medicines. Whether you're a cattle herder in Kenya, a duck farmer in Indonesia, or a dog owner in Mexico, your animals have benefited from the inexorably rising tide of modern veterinary medicine. And you've benefited as well.

Not only has modern veterinary medicine expanded its geographic reach, but, as the list of career paths my own classmates have taken indicates, it has also expanded its reach in other ways. Whereas Claude Bourgelat's first veterinary school only graduated veterinarians qualified to treat horses and cattle, today there is no species that does not have a veterinarian able to attend to it, and there is no animal-related activity that does not have a veterinarian involved somewhere.

And that is how it should be.

Epilogue

"What's his name?" I asked, as I gently lifted the puppy's lips to look at his gums. The puppy was barely moving, but his eyes flickered towards me. Normally I would know the name from the file, but the puppy had been rushed in and needed to be looked at before any particulars could be taken.

"Sirius," the man said. He was about my age. He wore a camo hat, and his T-shirt advertised something hunting related.

"The dog star," I said, continuing my examination.

The man nodded and cleared his throat. "So, what do you think? Is he going to be okay?"

"It's too soon to say for sure, but I hope so. How long have you had him?"

"Two days."

"From up north?"

"Yeah. Buddy of mine found him."

I nodded while waiting for the thermometer to register a reading: 39.7. A fever. There was bloody diarrhea on the thermometer. And the puppy had apparently been vomiting all day. I rarely saw these anymore because the vaccines are so effective and so widely used in the city. But if there ever was a textbook case, this was it.

I looked up at the man. "We'll have to do a test, but I think Sirius might have parvo."

"Is that bad?"

"Honestly, it's not good, but he's not that dehydrated yet, and you said he just started getting sick this morning, so he's got a decent chance of pulling through. He may need a plasma transfusion, though."

"Well, do whatever you need to. He's already a member of the family." The man chuckled and reached over to scratch Sirius behind the ears. There was a faint tail wag.

"I understand. We'll do our best for Sirius."

The man sighed and shook his head. "I grew up on the farm. Our dogs there never saw the vet. But with Buster, our last dog, we even took him to Saskatoon for cancer treatment. Cost five grand. But it bought him another good year, and it was totally worth it. I don't regret any of it. So as long as it isn't *ten* grand . . ." The man laughed.

At the sound of his laugh, the wag got a little stronger.

And there it is. This man, whose name I have forgotten (and which I would have changed anyway), embodies the story of veterinary medicine. In the span of a single generation, we've gone from farm dogs never being treated to routine chemotherapy and plasma transfusions. Although the power of science has made our medical tools infinitely more effective in the 14,000 years since the beginning of this journey, our motivations for doing so have come full circle. It was not the exact same virus, but otherwise there is strong resonance between Bonnie, the puppy who was treated for distemper by the old woman in prehistoric Germany, and Sirius, the puppy who was going to be treated for parvo by me. (Incidentally, you'll be happy to know that Sirius is doing very well now.)

Veterinary medicine has taken a long meandering road to arrive at a place once again where animals are integrated into our lives and are given medical care primarily because we love them and, if truth be told, because we value them in a way that could sometimes be described as spiritual. Granted, the word "spiritual" means different things to different people. While it is unlikely that we will begin to view dogs as holy the way the Zoroastrians did,* we are increasingly recognizing that animals have worth as individuals. We are acknowledging their similarities to us, and how human and animal life is intertwined.

Of course, thousands of veterinarians still treat animals entirely for practical reasons, and a large part of the world's food supply is dependent on this critical work. As I've mentioned before, the profession is increasingly split. An article in *Vox* earlier this year even called tension between animal rights–oriented veterinarians and their colleagues working in the meat industry a "civil war."** To this I would say that there is still more that binds us than divides us. The vast majority of all veterinarians base their decisions on concern for the welfare of animals. It's how that concern manifests, and how welfare is defined, that causes debate. Regardless, however, the trendline towards more care, more empathy, and a broader definition of animal welfare is clear and, in my opinion, unstoppable. It just may take some sectors of society longer to get there.

---

* It is interesting to note, however, that Pope Francis recently hinted that animals might be able to go to heaven.

** Marina Bolotnikova, "The Bitter Civil War Dividing American Veterinarians: To Fight the Cruel Meat Industry, Veterinarians Have to Fight Their Own Profession," *Vox*, Jan 4, 2023, https://vox.com/future-perfect/23516639/veterinarians-avma-factory-farming-ventilation-shutdown.

Acknowledgements

I have difficulty imaging how I would have written this book without access to *Veterinary Medicine: An Illustrated History* by Robert H. Dunlop and David J. Williams (Mosby, 1996). Clocking in at 707 pages and 3.4 kilograms (7.5 pounds) and bristling with illustrations, this is the definitive history of veterinary medicine. Unfortunately, it is out of print, and it was not written with the general reader in mind. It is dense. But for a veterinarian wanting to write about the history of his profession, it was indispensable as a road map. It provided me with the names and events and places to research. It would have been difficult to know where to begin otherwise — a matter of not even knowing what I didn't know. With a handful of exceptions, the historical veterinarians I featured are not household names, even within the profession. Dunlop and Williams made it clear what information I lacked and now needed to learn more about.

I also have difficulty imagining how I would have written this book without access to the internet. I'm not being flippant, nor do I mean that I used Quora and Reddit. I mean that the internet now allows you to view and read primary sources that once were the preserve of researchers with entrée into the British Library, the Bodleian Library, or the Library of Congress. With about 20 seconds of typing and clicking, and Delabere Blaine's 1817 *Canine Pathology* is in front of me here in Winnipeg, Canada. No cumbersome trans-Atlantic travel required. No discouraging applications for funding to pay for such travel required. I could not have written this book, at least the way I wrote it, before these public domain works became accessible online.

As much as possible, I used those primary sources as my principal references. Starting in ancient times, even as far back as those two lines in the Code of Hammurabi, the voice of the past speaks most vividly when allowed to speak for itself. This will have been most obvious to you in places where I extracted direct quotes or referred to them in the discussion; they also underpinned most of the short stories, or vignettes, that opened each chapter. It goes without saying that I had to make use of my imagination to conjure specific conversations or some of the fine detail, but the primary sources informed all the important factual matters conveyed in these stories. Professional historians will cringe at the conjecture inherent in employing fiction writing techniques in a book purporting to tell the truth about the past. To that cringe, my response is that I put considerable effort into making any conjecture plausible, even highly plausible, and that what these stories have to say about the state of the veterinary arts at the time is accurate. At least as accurate as anything written hundreds or thousands of years after the fact can be.

I also want to take advantage of this space to thank the marvellous people at ECW Press in Toronto. They have been nothing but supportive and helpful. And Jack David deserves extra thanks for taking a chance on what is admittedly an unusual concept.

Sources and Further Reading

CHAPTER TWO: THE OLD WOMAN

Alex, Bridget. "Prehistoric Medicine: How Archaic Humans Cured Themselves." *Discover*. May 10, 2019. https:/discovermagazine.com/planet-earth/prehistoric-medicine-how-archaic-humans-cured-themselves.

Coren, Stanley. "Is Emotional Attachment to Dogs a Modern Development?" *Psychology Today*. February 14, 2018. https://psychologytoday.com/ca/blog/canine-corner/201802/is-emotional-attachment-dogs-modern-development.

Janssens, Luc, Liane Giemsch, Ralf Schmitz, Martin Street, Stefan Van Dongen, and Philippe Crombé. "A New Look at an Old Dog: Bonn-Oberkassel Reconsidered." *Journal of Archaeological Science* 92 (2018): 126–138. https://doi.org/10.1016/j.jas.2018.01.004.

CHAPTER THREE: UBAR

Mark, Joshua J. "A Brief History of Veterinary Medicine." World History Encyclopedia. April 30, 2020. https://worldhistory.org/article/1549/a-brief-history-of-veterinary-medicine/.

Mark, Joshua J. "Dogs in Ancient Persia." World History Encyclopedia. December 5, 2019. https://worldhistory.org/article/1482/dogs-in-ancient-persia/.

CHAPTER FOUR: PALAKAPYA

Rutkow, Ira. *Empire of the Scalpel: The History of Surgery*. New York: Scribner, 2022.

Sheshadri, K.G. "*Basti* Therapy of Elephants According to Sage Palakapya." *Gajah* 43 (2015): 36–41. https://asesg.org/PDFfiles/2015/Gajah%2043/43-36-Sheshadri.pdf.

CHAPTER FIVE: VEGETIUS

Renatus, Vegetius. *Of the Distempers of Horses and of the Art of Curing Them*. English translation of *Digesta Artis Mulomedicinae*. London: A. Millar, 1748. https://vethistory.rcvsknowledge.org/vegetius-renatus-of-the-distemper-of-horses-and-of-the-art-of-curing-them-as-also-of-the-diseases-of-oxen-and-of-the-remedies-proper-for-them-and-of-the-best-method-to-preserve/.

CHAPTER SIX: OSWALD

Electric Scotland. "Scottish Charms and Amulets: Elf-Arrows." https://electricscotland.com/history/articles/charms6.htm.

Electric Scotland. "Scottish Charms and Amulets: The Lee-Penny." https://electricscotland.com/history/articles/charms13.htm.

Medieval Manuscripts (blog). "Bald's Leechbook Now Online." British Library. January 5, 2016. https://blogs.bl.uk/digitisedmanuscripts/2016/01/balds-leechbook-now-online.html.

CHAPTER SEVEN: FÉBUS

Edward of Norwich, second Duke of York. *The Master of Game*, c. 1406–1413. English translation of Gaston Fébus's *Livre de chasse*, c. 1387–1389. Edited by William A. Baillie-Grohman and F. Baillie-Grohman. London: Chatto & Windus, 1909. https://ia903100.us.archive.org/0/items/themasterofgamet43452gut/43452-h/43452-h.htm.

Snape, Andrew. *The Anatomy of an Horse*. London: Miles Fletcher for Thomas Fletcher, 1683.

CHAPTER EIGHT: FITZHERBERT

Fitzherbert, Anthony. *The Book of Husbandry*. Edited by Walter W. Skeat. London: Trübner & Co., [1534] 1882. https://gutenberg.org/files/57457/57457-h/57457-h.htm.

Markham, Gervase. *Markham's Maister-Peece*. London: Nicholas Okes, 1610. In the digital collection *Early English Books Online*. University of Michigan Library Digital Collections. https://name.umdl.umich.edu/A06950.0001.001.

CHAPTER NINE: LANCISI

Mantovani, A., and R. Zanetti. "Giovanni Maria Lancisi: De Bovilla Peste and Stamping Out." *Historia Medicinae Veterinariae* 18, no. 4 (1993): 97–110. https://pubmed.ncbi.nlm.nih.gov/11639894/.

Roeder, Peter, Jeffrey Mariner, and Richard Kock. "Rinderpest: The Veterinary Perspective on Eradication." *Philosophical Transactions of the Royal Society of London: Series B, Biological Sciences* 358, no. 1623 (2015): 20120139. https://ncbi.nlm.nih.gov/pmc/articles/PMC3720037/.

CHAPTER TEN: SINOPA

Claymore, Sunshine. "Ethnoveterinary Treatment of Horses by the Great Sioux Nation." PowerPoint presented at the Missouri Botanical Garden, St. Louis, MO, November 16, 2012. https://

missouribotanicalgarden.org/Portals/0/Science%20and%20Conservation/PDFs/REU/2012/Claymore-2012-REU-%20Presentation.pdf.

Lans, Cheryl. "Possible Similarities between the Folk Medicine Historically Used by First Nations and American Indians in North America and the Ethnoveterinary Knowledge Currently Used in British Columbia, Canada." *Journal of Ethnopharmacology* 192 (November 4, 2016): 53–66. https://doi.org/10.1016/j.jep.2016.07.004.

Pyramid Mesa (blog). "The Orphan Boy and the Elk Dog." November 13, 2000. https://web.archive.org/web/20121113125104/http://www.pyramidmesa.com/blackfoot1.htm.

CHAPTER ELEVEN: BOURGELAT

Larkin, Malinda. "Pioneering a Profession: The Birth of Veterinary Education in the Age of Enlightenment." American Veterinary Medical Association. December 19, 2010. https://avma.org/javma-news/2011-01-01/pioneering-profession.

CHAPTER TWELVE: CLARK

Clark, James. *A Treatise on the Prevention of Diseases Incident to Horses from Bad Management in Regard to Stables, Food, Water, and Exercise. To which are Subjoined Observations on some of the Surgical and Medical Branches of Farriery*. Edinburgh: Alex Smellie, 1802. https://books.google.ca/books?id=FqXhLib3lKYC.

CHAPTER THIRTEEN: ERXLEBEN

Dennis, Jerry. "A History of Captive Birds." *Michigan Quarterly Review* 53, no. 3 (summer 2014). http://hdl.handle.net/2027/spo.act2080.0053.301.

Erxleben, Johann Christian Polycarp. *Theoretischer Unterricht in der Vieharzneykunst*. Gottingen: Johann Christian Dieterich, 1798. https://wellcomecollection.org/works/jkecnwsw/items?canvas=17.

Howard-Smith, Stephanie. "Mad Dogs, Sad Dogs, and the 'War against Curs' in London in 1760." *Journal for Eighteenth-Century Studies* 42, no. 1 (March 2019): 101–118. https://doi.org/10.1111/1754-0208.12592.

CHAPTER FOURTEEN: MOORCROFT

Alder, Garry. *Beyond Bokhara: The Life of William Moorcroft, Asian Explorer and Pioneer Veterinary Surgeon, 1767–1825*. Delhi: Low Price Publications, 2012. https://archive.org/details/dli.pahar.3779.

Rashid, Salman. "A Forgotten Page from History." *Odysseus Lahori* (blog). September 7, 2013. https://odysseuslahori.blogspot.com/2013/09/ShalamarGarden.html?m=1.

CHAPTER FIFTEEN: BLAINE

Blaine, Delabere Pritchett. *Canine Pathology*. London: T. & T. Boosey, 1832. https://archive.org/details/canine-pathology.

CHAPTER SIXTEEN: SMITH

Barker, C.A.V. "The Ontario Veterinary College: Temperance Street Era." *The Canadian Veterinary Journal* 16, no. 11 (November 1975): 319–328. https://ncbi.nlm.nih.gov/pmc/articles/PMC1697073/pdf/canvetj00408-0005.pdf.

Evans, A.M. "Smith, Andrew." In *Dictionary of Canadian Biography*, Volume 13, online. http://biographi.ca/en/bio/smith_andrew_13E.html.

CHAPTER SEVENTEEN: GALTIER

Cavaillon, Jean-Marc, and Sandra Legout. "Louis Pasteur: Between Myth and Reality." *Biomolecules* 12, no. 4 (2022): 596. https://doi.org/10.3390/biom12040596.

Theodorides, J. "Pasteur and Rabies: The British Connection." *Journal of the Royal Society of Medicine* 82 (August 1989): 488–490. https://doi.org/10.1177/014107688908200813.

CHAPTER EIGHTEEN: MCGRATH

Katić, Ivan. "Pioneer Female Veterinarians." *Medical Sciences* 37 (2012): 137–168. https://hrcak.srce.hr/file/125056.

O'Brien, Lora. "Ireland's First Female Veterinarian." *Lora O'Brien — Irish Author & Guide* (blog). July 4, 2018. https://loraobrien.ie/irelands-first-female-veterinarian/.

CHAPTER NINETEEN: COLEBOURN

Canadian Great War Project. "Major Harry Colebourn." Last updated January 30, 2014. https://canadiangreatwarproject.com/person.php?pid=69285.

Shushkewich, Val. *The Real Winnie: A One-of-a-Kind Bear.* Toronto: Dundurn Press, 2005.

CHAPTER TWENTY: WIGHT

Jones, David. "'Dad's Vet Books Brought Joy to So Many . . . Yet He Lived with an Aching Sadness': Children of *All Creatures Great and Small* Author James Herriot Reveal the Truth behind Novels That Made the Queen Laugh Out Loud." *Daily Mail.* September 23, 2020. https://dailymail.co.uk/femail/article-8765955/James-Herriots-children-reveal-truth-Creatures-Great-Small-series-lifts-spirits.html.

Natterson-Horowitz, Barbara, and Kathryn Bowers. *Zoobiquity: The Astonishing Connection between Human and Animal Health*. New York: Vintage, 2013.

Wight, Jim. *The Real James Herriot: A Memoir of My Father*. New York: Ballantine Books, 2001.

CHAPTER TWENTY-ONE: CAMUTI

Camuti, Louis J. *All My Patients Are Under the Bed: Memoirs of a Cat Doctor*. New York: Simon and Shuster, 1980.

CHAPTER TWENTY-TWO: BLOOD

Blood, D.C., and J.A. Henderson. *Veterinary Medicine*. London: Bailliere, Tindall & Cox, 1963.

CHAPTER TWENTY-THREE: KALEMA-ZIKUSOKA

Kalema-Zikusoka, Gladys. *Walking with Gorillas: The Journey of an African Wildlife Vet*. New York: Arcade, 2023.

National Georgaphic Expeditions. "Gladys Kalema-Zikusoka, Conservationist." https://nationalgeographic.com/expeditions/experts/gladys-kalema-zikusoka/.

Schott, Philipp. *How to Examine a Wolverine: More Tales from the Accidental Veterinarian*. Toronto: ECW Press, 2021.

Entertainment. Writing. Culture.

ECW is a proudly independent, Canadian-owned book publisher. We know great writing can improve people's lives, and we're passionate about sharing original, exciting, and insightful writing across genres.

Thanks for reading along!

We want our books not just to sustain our imaginations, but to help construct a healthier, more just world, and so we've become a certified B Corporation, meaning we meet a high standard of social and environmental responsibility — and we're going to keep aiming higher. We believe books can drive change, but the way we make them can too.

Being a B Corp means that the act of publishing this book should be a force for good — for the planet, for our communities, and for the people that worked to make this book. For example, everyone who worked on this book was paid at least a living wage. You can learn more at the Ontario Living Wage Network.

This book is also available as a Global Certified Accessible™ (GCA) ebook. ECW Press's ebooks are screen reader friendly and are built to meet the needs of those who are unable to read standard print due to blindness, low vision, dyslexia, or a physical disability.

The interior of this book is printed on Sustana EnviroBook™, which is made from 100% recycled fibres and processed chlorine-free.

ECW's office is situated on land that was the traditional territory of many nations, including the Wendat, the Anishinaabeg, Haudenosaunee, Chippewa, Métis, and current treaty holders the Mississaugas of the Credit. In the 1880s, the land was developed as part of a growing community around St. Matthew's Anglican and other churches. Starting in the 1950s, our neighbourhood was transformed by immigrants fleeing the Vietnam War and Chinese Canadians dispossessed by the building of Nathan Phillips Square and the subsequent rise in real estate value in other Chinatowns. We are grateful to those who cared for the land before us and are proud to be working amidst this mix of cultures.

ecwpress.com